THE OXIDATION-REDUCTION POTENTIAL IN GEOLOGY

PROBLEMA OKISLITEL'NO-VOSSTANOVITEL'NOGO POTENTSIALA V GEOLOGII

ПРОБЛЕМА ОКИСЛИТЕЛЬНО-ВОССТАНОВИТЕЛЬНОГО ПОТЕНЦИАЛА В ГЕОЛОГИИ

THE OXIDATION-REDUCTION POTENTIAL IN GEOLOGY

Mikhail F. Stashchuk

Director, Laboratory of Mineral Forming Processes
Institute of Mineral Resources
Ministry of Geology
Simferopol, Ukrainian SSR

Translated from Russian by
J. Paul Fitzsimmons
Professor of Geology
University of New Mexico
Albuquerque, New Mexico

(c/b) CONSULTANTS BUREAU • NEW YORK—LONDON • 1972

Mikhail Fedorovich Stashchuk was born May 7, 1931, in Kiev. He was graduated from Chernovtsy University in 1954. From 1954 to 1957 he was a candidate at the Institute of Geological Sciences of the Academy of Sciences of the Ukrainian SSR, and in 1958 he defended his dissertation at Kiev University. Upon finishing his graduate studies, he was employed at the Institute of Mineral Resources in Simferopol. In 1963, by resolution of the Presidium of the Academy of Sciences of the Ukrainian SSR, the scientific title of Senior Scientist with Specialty in Mineralogy and Petrography was conferred upon him. At present he is director of the Laboratory of Mineral-Forming Processes at the Institute of Mineral Resources of Geology, Ukrainian SSR, Simferopol.

The Russian text, originally published by Nedra Press in Moscow in 1968, has been revised and corrected by the author for this edition. The present translation is published under an agreement with Mezhdunarodnaya Kniga, the Soviet book export agency.

Михаил Федорович Стащук

ПРОБЛЕМА ОКИСЛИТЕЛЬНО-ВОССТАНОВИТЕЛЬНОГО ПОТЕНЦИАЛА В ГЕОЛОГИИ

Library of Congress Catalog Card Number 75-37615
ISBN 978-1-4684-1595-7   ISBN 978-1-4684-1593-3 (eBook)
DOI 10.1007/978-1-4684-1593-3

© 1972 Consultants Bureau, New York
A Division of Plenum Publishing Corporation
227 West 17th Street, New York, N. Y. 10011

United Kingdom edition published by Consultants Bureau, London
A Division of Plenum Publishing Company, Ltd.
Davis House (4th Floor), 8 Scrubs Lane, Harlesden, London, NW10 6SE, England

PREFACE TO THE AMERICAN EDITION

Within very recent time, when investigating the physicochemical conditions under which sedimentary rocks formed, geologists were satisfied with logical premises prompted by sound judgment. Recognizing the great role of this factor in development of the science, it is still necessary to keep in mind that sound judgment based on impressions and subjective experience of the observer does not always bring us to an understanding of objective reality.

Probably greater success in explaining the physicochemical features of the environment in which sediments accumulate may be achieved by mathematical treatment on the basis of fundamental laws of nature. It is true that such treatment, as in solutions of ordinary arithmetic problems, frequently yields several formally true answers, only one of which satisfies actual natural processes. Unrestricted reference to thermodynamic computations may therefore lead to serious errors and may instill mistrust of this progressive trend among geologists.

This warning seems to me to be in no way exaggerated, since we have noted in recent years a strong tendency to introduce chemical thermodynamics into lithology. The impetus for this was given by works of American geochemists using a graphical method, permitting easy and clear representation of interrelations between minerals and allowing one to explain possible transformations according to variations in different characteristics of the environment.

The present book is written within the indicated framework. I am pleased that this book, in which the graphical method is used and popularized, has been translated into English.

In giving a physicochemical evaluation of existing methods for determining the oxidation − reduction conditions under which sedimentary rocks form, I have tried to present the material in such a way that the warning against too formal an application of thermodynamics is obvious.

No fundamental changes have been introduced into this edition. Unfortunate words, expressions, and misprints have been corrected. The symbol G has been introduced in place of the Z used to designate isobaric − isothermal potential in the Soviet edition. The symbol Z has been frequently encountered in science, chiefly in the geologic literature. (In the American literature, the letter F is used just as frequently.) In Soviet and foreign chemical literature, this parameter of state has been designated by G in recent years in honor of the American scientist Josiah Willard Gibbs, who laid the foundation for analyzing processes by means of isobaric − isothermal potential.

Simferopol'   M. F. Stashchuk

PREFACE

In solving practical problems and establishing a firm basis for theoretical discussions, geologists are faced with the fundamental task of explaining the conditions under which rocks form. Various concepts are adopted as bases for geological surveying, prospecting, and exploration for mineral deposits. Throughout all the stages of geologic investigations these are developed and refined. Proper genetic views free the geologist from unnecessary and unproductive work. For geologists having to do with sedimentary rocks, tremendous significance is found in the investigation of conditions under which sediments accumulate and under which these sediments are altered and converted to rock. A large number of investigators have devoted their labors to these questions, and their ideas form the foundation of the theory of sedimentary ore formation. Special attention in this respect belongs to the papers of A. E. Fersman, V. I. Vernadskii, N. M. Strakhov, M. S. Shvetsov, L. V. Pustovalov, W. H. Twenhofel, L. B. Rukhin, and others.

When investigating the conditions under which sedimentary rocks are formed, a large role must be ascribed to the physicochemical processes accompanying the accumulation of sediments and diagenesis. This in turn presupposes a certain oxidation-reduction potential and acid-alkali regime at the time the sedimentary rocks were forming. Because of the great significance of these parameters, many papers have appeared in recent years in connection with the application of Eh and pH and with methods of determining them. We can hardly fail to notice, however, that, at the same time, serious discrepancies began to be noticed, both in the conception of oxidation-reduction potential itself and in the methods of determining it.

The discrepancies that have appeared and are appearing cannot be resolved without a clear answer to the following question: what is included in the concept of oxidation-reduction conditions accompanying individual stages in the formation of sedi-

mentary rocks? What method furnishes a true oxidation-reduction characteristic? Can we, in general, find such a method, or must we, at the present stage, restrict ourselves to a quantitative comparison of strata and to an approximation of physicochemical calculations? The present book is devoted to an investigation of these questions.

The work of one of the leading geochemists, R. M. Garrels, has had great influence on the author. Garrels' work is responsible for the fact that this book considers the problem from the physicochemical viewpoint and uses the methods of thermodynamic computations employed by Garrels. It should be noted that a critical approach to the Garrels' diagrams would not have been possible without the detailed investigations of P. A. Kryukov and V. M. Levchenko on the system of iron and sulfur, carried out by them since the thirties.

Since this book is intended for the widest circle of readers, including those unacquainted with the principles of thermodynamics, we have prefaced the necessary basic part of the investigation with a detailed discussion of the physicochemical viewpoint used in the book. Strict mathematical proofs of particular laws requiring definite knowledge of either higher mathematics or physical chemistry are replaced by examples and clear graphical material so far as is possible. The discussion, popular at first, gradually becomes more rigorous, but does not require any new mathematical technique. The author has long studied both the oldest deposits (Archean) and the youngest (Quaternary loams and Holocene lagoonal and lacustrine deposits) and may therefore be considered well equipped to furnish examples from his own experience. In order to avoid subjectivism, however, proofs of the different viewpoints are taken from already published data, from various investigators, modified to some extent for unity and clarity of presentation but with this modification always specified in the text. In general, we have tried not to overload the book with geologic examples, believing that the reader may find confirmation or contradiction of the investigated problems in the pursuit of his personal research.

   The basic problem is investigated in its application to iron minerals, the most widespread authigenic minerals in sedimentary rocks. As an example, manganese minerals are investigated. The author hopes, however, that the method used and the factual material may serve to explain the conditions under which any other authigenic minerals are formed.

   The author expresses his sincere thanks to S. K. Kropacheva and V. M. Lobanova-Sopit'ko, who aided him in the computations, F. I. Stashchuk, whom the author consulted on certain questions in discussions of physical phenomena, and M. D. Stashchuk and I. Ya. Neipak, who prepared the manuscript for printing.

   A special feeling of gratitude is felt by the author toward A. I. Perel'man and Yu. Yu. Yurk, without whose active support the work could hardly have been completed. The manuscript was read thoroughly by V. V. Shcherbina, who made a whole series of constructive remarks for its improvement and who proved to be of invaluable service to the author.

CONTENTS

CONTENTS

CHAPTER 1

THE PRINCIPAL METHODS OF DETERMINING OXIDATION—REDUCTION CONDITIONS

Many chemical elements have the peculiar attribute, depending on existing conditions, of changing their valence and, with this, their properties. For example, iron may have zero valence or may be divalent or trivalent; manganese may be 0, +2, + 3, +4, and more; sulfur may be −2, 0, +6; and so forth. The remarkable change in properties of chemical elements with change in valence determines one of the decisive factors in the migration and concentration of these elements. For example, the expressed features of alkalinity in bivalent manganese and iron lend these elements a broader migrational capacity in the sedimentary shell of the earth than tervalent iron or quadrivalent manganese. Elemental sulfur is many times less mobile than its oxidized or reduced forms. The great significance of valence change in the migration and concentration of chemical elements was pointed out by Goldschmidt [1933], Fersman [1955], Shcherbina [1939, 1949, 1956, 1965], Pustovalov [1940], Perel'man [1965], and others. Fersman [1955] noted that the oxidation−reduction potential has a very special meaning in explaining the order of coprecipitation in the supergene environment. On the basis of normal potentials, Shcherbina [1939] divided the elements of different valences into groups, one of which manifested a tendency to coprecipitate with the ferrous ion, the other with the ferric. Elements that are concentrated under oxidizing conditions are disseminated under reducing conditions, and the reverse [Shcherbina, 1949, 1956]. Specific features manifested in trace elements during valence changes lead to concentrations chiefly under strongly oxidizing conditions or under strongly reducing conditions [Mason, 1949]. It is no exaggeration to state that without knowledge of the oxidation−reduction state during the process of sedimentary rock formation one cannot seriously approach a solution to the problems of concentration and dissemination of elements in the sedimentary shell of the earth.

The oxidation−reduction potential has great significance in prospecting for oil deposits [Levenson, 1964] and for deposits of elemental sulfur. In these cases it most frequently determines the secondary processes that develop through the influences of the oil deposit, becoming an indicator of the "dissemination aureole" of a mineral deposit. These examples show the importance of studying those criteria that may reliably characterize the oxidation−reduction state of sedimentary rocks in the process of forming and the amount of superposition of secondary oxidation−reduction processes even where the rock has not been subjected to the epigenetic influence of atmospheric oxygen.

A solution of such problems may not be reached without inclusion of the basic concepts of physical chemistry, on which our ideas of the oxidation−reduction potential are based. In this connection, as early as 1933, Pustovalov wrote that "the geologist must possess geochemical and chemical knowledge to such an extent that in his observations of nature he may in his imagination reconstruct the physicochemical conditions under which any particular chemical reaction took place" [1933a, p. 23].

Unfortunately, the tendency toward physicochemical interpretation of conditions under which sedimentary rocks form sometimes grows more rapidly than the preparedness of the specialists in the field. In the geologic literature of the present time, many inaccurate concepts are therefore accumulating, and are finding wide application in various theories concerning the conditions under which sedimentary rocks form. We may even cite some examples: when an acidic environment and oxidizing conditions are taken as synonymous [Babaev, 1957a, 1957b] or when values of Eh of +40 and +56.2 mV are ascribed to a weakly reducing environment while nearby a value of +56 mV is considered representative of a strongly reducing environment [Ismailov, 1959].

It would be too impractical to analyze these cases, since they very clearly represent errors. But there exists a whole series of opinions that appear outwardly convincing though they have no proper physicochemical basis. In accumulating, one after the other, they eventually lead to an orderly theory, the erroneous nature of which is difficult to detect. Such errors, in our view, represent the principal reasons for the appearance of numerous mutually inconsistent methods of determining the oxidation–reduction conditions under which sedimentary rocks form. Let us briefly examine these methods.

In the opinion of a number of investigators, the oxidation–reduction conditions during the formation of sedimentary rocks may be characterized by the following parameters:

1. The ratio of ferrous to ferric forms of iron in the sedimentary rock. Two variations of this method exist: a) the ratio of the forms of iron determined in making bulk analysis [Ronov, 1958], and b) the ratio of the mobile forms of iron [Klenova, 1933; Romm, 1950; and others);

2. Quantitative determination of the different forms of sulfur;

3. The quantity of residual organic matter;

4. The group of iron minerals and their paragenetic relations. This method may be treated in two ways, leading to fundamentally different conclusions: Pustovalov [1933b] and Strakhov [Strakhov and Zalmanzon, 1955];

5. Direct electrometrical measurements on the rocks.

The $Fe^{..}/Fe^{...}$ Ratio and Eh

The method involving the ratio of the forms of iron, used as an indicator of the oxidation–reduction conditions, is the oldest. It was based on the fact that the change in valence during conversion from Fe_2O_3 to FeO must take place as a consequence of diminution in content of oxygen, and this, naturally, characterizes an increase in the reducing capacity of the environment. The method was proposed by V. M. Goldschmidt, and is presently used by Ronov [1958]. As several investigators have noted, individual variations may arise with this method as a consequence of the fact that the ratio of the forms of iron due to specific formation in the sediments will be modified by admixtures of ferric and ferrous iron falling into the basin as clastic material and not being subjected to alteration in the muds.

The necessity of getting away from the ballast that is inert to the oxidation–reduction process has led to the search for methods of distinguishing "mobile" forms of iron, i.e., those forms that are not present in sediment of clastic origin. Strictly speaking, the basis of the idea of analyzing mobile forms of iron appeared much earlier, when Schmelck [1882] suggested that the color of sediments be assigned by the relations of Fe_2O_3 and FeO, but it was not then felt necessary to distinguish the forms of iron by accurate methods, since the quantitative value (the Fe_2O_3/FeO ratio) yielded an evaluation of qualitative characteristic (color of the sediments).

However, as soon as the idea appeared that some kind of relationship must exist between the Fe_2O_3/FeO ratio and the content of oxygen in the environment, the amount of organic matter, or the oxidation–reduction potential (Eh), it became urgent to determine precisely the mobile forms of iron, which would make it possible to clarify the functional relations connecting these values.

In 1933 Klenova suggested a special method of determining the forms of iron in a 10% hydrochloric-acid solution immediately after collection of the sample from the floor of the basin. The first correlations, made by Klenova [1938] on sediments of the Motovskii Gulf, showed that the connection between the ferric–ferrous iron ratio and the oxygen content in the muds is not very clear (Fig. 1). Klenova, clearly assuming that the results obtained represented a first attempt at correlation, and basing her conclusions more on a qualitative comparison than on a quantitative evaluation, concluded, somewhat hastily in our opinion, that this ratio ($Fe^{...}/Fe^{..}$) represents a real reflection of the oxidation–reduction potential.

With difficulty, controversial logical grounds may be found to support this conclusion. The first unsuccessful correlation was therefore not abandoned, and the search for systematic correlation between the ratio of forms of iron and the oxidation–reduction potential was continued. The meth-

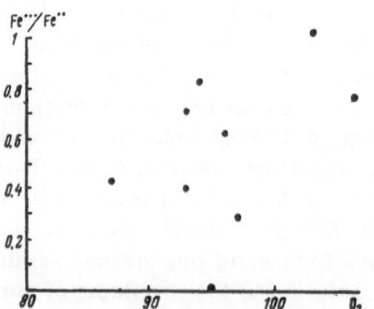

Fig. 1. Relations of forms of iron and oxygen in recent sediments (prepared from the data of Klenova [1938]).

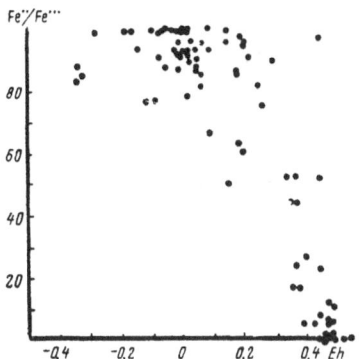

Fig. 2. Nature of the relation between iron ratio and oxidation–reduction potential in muds from the northeastern part of the Pacific Ocean (prepared from the data of Romankevich and Petrov [1961].

od of analysis was altered somewhat, the ratio Fe$^{\cdot\cdot\cdot}$/Fe$^{\cdot\cdot}$ being replaced by Fe$^{\cdot\cdot}$/Fe$^{\cdot\cdot\cdot}$, and the correlation itself was made not relative to oxygen but relative to Eh or rH$_2$. Instead of 10% hydrochloric acid, 5% acid was used [Romm, 1950], in order not to touch pyritic iron, which, as was assumed, did not affect the oxidation–reduction potential of the environment. Later, in order to improve the reliability of obtaining just the mobile forms of iron, a 2% solution of hydrochloric acid was used instead of 5% and the time of boiling the sample was reduced [Strakhov and Zalmanzon, 1955].

Analyses of various sediments, carried out by these methods, apparently began to give reassuring results. This is indicated particularly by the data of Romankevich and Petrov [1961] on bottom sediments from the northeastern part of the Pacific Ocean, which we have plotted in a diagram (Fig. 2). As seen from this figure, the tendency for Eh to decline with increase in the Fe$^{\cdot\cdot}$/Fe$^{\cdot\cdot\cdot}$ ratio stands

out more or less clearly. However, new data, obtained from the Indian Ocean by Zheleznova and Shishkina [1964] and Isaeva [1964], shown in Fig. 3, again raise some doubt that there must exist some clear correlation, stochastic if not functional, between the Fe$^{\cdot\cdot}$/Fe$^{\cdot\cdot\cdot}$ ratio and the oxidation–reduction potential.

A careful examination of Fig. 3 leads one to suggest that here the following feature is manifested: the points characterizing the Fe$^{\cdot\cdot}$/Fe$^{\cdot\cdot\cdot}$ ratio occupy two different fields relative to Eh. One field lies to the left of Eh = 0.1 V. Within this field the points exhibit considerable scatter, relative both to the Eh value and to the Fe$^{\cdot\cdot}$/Fe$^{\cdot\cdot\cdot}$ ratio. These two values do not correlate between themselves. A well defined field lies to the right of Eh $^-$ 0.1 V. In this field the points press close to the abscissa axis, which only emphasizes independence between the Fe$^{\cdot\cdot}$/Fe$^{\cdot\cdot\cdot}$ ratio and Eh.

When we compare the distribution of points with the characteristics of the analyzed sediments (Table 1), it becomes perfectly clear that the left-hand field characterizes muds in which hydrotroilite forms, the right-hand field by muds in which the iron mineral is some hydroxide. A similar picture is observed for muds of the Pacific Ocean, the characteristics of which are shown in Fig. 2: to the left of Eh $\approx$ 0.3 V are found gray muds in which hydrotroilite occurs, but to the right of Eh $\approx$ 0.3 V are found brownish muds with iron hydroxides.

May we expect to find a dependent relationship between the iron ratio and the oxidation–reduction potential within each of these fields? A single glance at Figs. 2 and 3 is sufficient to convince ourselves that no well-defined correlation between the investigated values is to be found within either field. Confirmation of this conclusion is found in data from basins in which sediments with monotypic mineralogical characteristics have accumulated. This is

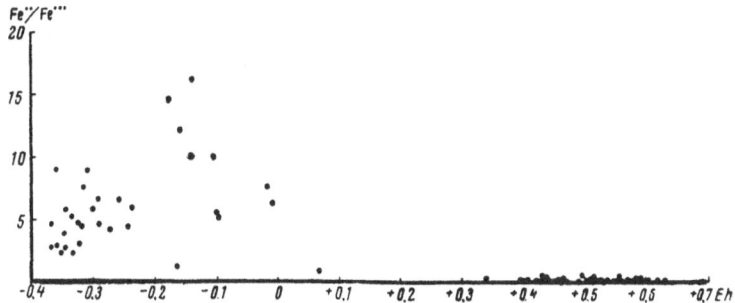

Fig. 3. Correlation between iron ratio and oxidation–reduction potential in muds from the northern part of the Indian Ocean (plotted from the data of Zheleznova and Shishkina [1964] and Isaeva [1964]).

TABLE 1. Oxidation–Reduction Potential, pH, and Iron Ratios in Sediments from the Northern Part of the Indian Ocean (after Zheleznova and Shishkina [1964] and Isaeva [1964])

Station No. and layer in core and dredge sample	Depth, m	Description of sediment	pH	Eh, mV	Fe$^{\cdot\cdot}$	Fe$^{\cdot\cdot\cdot}$	Fe$^{\cdot\cdot}$/Fe$^{\cdot\cdot\cdot}$
		Arabian Sea					
4798	4365						
14—40		Mud, pale yellow, effervesces in HCl	—	+518	—	—	—
51—70		Mud, straw-colored, dense, viscous, homogeneous, effervesces in HCl	—	+501	—	—	—
96—115		The same	—	+491	—	—	—
149—170		The same, with concretions	—	+596	—	—	—
196—213		Mud, grayish brown, effervesces in HCl	—	+544	—	—	—
Upper 4799 part of dredge sample	4458	Mud, calcareous, argillaceous, grayish brown with dark layers	7.42	+515	0.08	0.20	0.40
Lower part		The same	7.43	+431	0.07	0.15	0.47
37—49		Mud, calcareous, silty-argillaceous, light gray with rare inclusions of hydrotroilite	7.42	−322	0.18	0.06	3.00
80—101		Mud, calcareous, argillaceous, light gray with rare inclusions of hydrotroilite	7.12	−315	0.30	0.04	7.50
190—204		Mud, calcareous, argillaceous, gray with greenish tone, many inclusions of hydrotroilite	−7.43	−323	0.24	0.05	4.80
290—308		The same	7.35	−229	0.40	0.04	10
Upper 4804 part of dredge sample	3717	Mud, semiliquid	7.44	−371	—	—	—
Lower part		Mud, greenish gray	7.39	−342	—	—	—
35—52		Mud, greenish gray	7.35	−307	—	—	—
85—100		The same, with small inclusions of hydrotroilite, odor of H$_2$S	7.35	−129	—	—	—
Upper 4807 part of dredge sample	758	Mud, greenish gray, weak odor of H$_2$S toward bottom	7.04	−212	—	—	—
Lower part		The same	6.99	−186	—	—	—
Upper 4808 part of dredge sample	3384	Mud, gray, semiliquid, fine brownish film at top	7.09	−352	—	—	—
Lower part		The same	7.19	−167	—	--	—
4809	2592						
10—20		Mud, gray, homogeneous	7.37	−205	—	—	—
109—120		Mud, gray	7.48	−231	—	—	—
249—270		Mud, gray, layers of silt from 265 level	7.41	−258	—	—	—
Upper 4812 part of dredge sample	471	Sand, fine, grayish green, effervesces in HCl	7.20	−157	—	—	—
Lower part		The same	7.22	−139	—	—	—
Upper 4814 part of dredge sample	3676	Mud, dark brown, effervesces in HCl	6.99	+424	Not det	0.33	0
Lower part		Mud, light brown, more dense	7.07	+408	0.05	0.27	0.19
30—49		Mud, greenish gray, effervesces in HCl	7.34	−366	0.34	0.12	2.83
140—160		The same, hydrotroilite present	7.46	−345	0.43	0.11	3.91
250—269		The same, amount of hydrotroilite increases	7.23	−290	0.33	0.07	4.71
350—376		The same	7.25	−287	0.49	0.12	4.08
4821	3874						
10—30		Mud, pale yellow, effervesces in HCl	7.29	+503	—	—	—
50—70		The same	7.45	+497	—	—	—
180—197		Mud, pale yellow with small dark grayish spots	7.42	+491	—	—	—

TABLE 1 continued

Station No. and layer in core and dredge sample	Depth, m	Description of sediment	pH	Eh, mV	Fe··	F···	Fe··/F···
280—298		Mud, pale yellow with small dark grayish spots	7.39	+498	—	—	—
380—394		The same	7.43	+517	—	—	—
Middle 4846 layer of dredge sample	3686	Mud, pale grayish yellow, effervesces in HCl	7.08	+503		—	
Upper 4848 part of dredge sample	4753	Mud, dark brown, effervesces weakly in HCl	7.31	+536	—	—	—
Lower part		The same, two concretions observed	7.30	+546	—	—	—
4848	4778						
35—53		Mud, dark brown, homogeneous, effervesces weakly in HCl	7.31	+535	—	—	—
85—110		The same	7.29	+507	—	—	—
130—150		" "	7.35	+518	—	—	--
158—178		Mud, gray, denser, effervesces in HCl	7.36	+24	—	—	—
Upper 4854 part of dredge sample	3172	Mud, calcareous, argillaceous, brown	7.42	+395	0.04	0.86	0.05
Lower part		Mud, calcareous, silty-argillaceous, light straw color	7.42	+339	0.07	0.35	0.20
30—42		The same, gray	7.36	+68	0.53	0.71	0.75
100—120		The same	7.29	—243	0.92	0.21	4.38
215—236		The same, hydrotroilite present	7.26	—308	0.99	0.11	9.00
300—317		Mud, calcareous, silty-argillaceous, gray, hydrotroilite present	7.33	—97	0.97	0.19	5.10
384—400		The same	7.31	—335	0.88	0.17	5.18
510—529		" "	7.31	—300	0.93	0.16	5.81
615—635		" "	7.33	—294	0.99	0.15	6.60
720—740		The same, very little hydrotroilite	7.34	—238	0.99	0.17	5.82

Indian Ocean

Station No. and layer in core and dredge sample	Depth, m	Description of sediment	pH	Eh, mV	Fe··	F···	Fe··/F···
4836	4392						
5—20		Mud, effervesces in HCl (pale yellow)	7.39	+519	Not det	0.27	0
40—55		The same, but darker	7.46	+536	» »	0.29	0
95—112		Mud, grayish brown, denser, effervesces in HCl	7.40	+553	» »	0.29	0
164—185		Mud, brown, effervesces in HCl	7.40	+550	» »	0.33	0
260—278		The same	7.40	+535	» »	0.28	0
4878	4676						
Middle layer of dredge sample		Mud, calcareous, argillaceous, brown, many manganese concretions	7.58	+576	» »	0.28	0
20—40		The same	7.76	+557	» »	0.33	0
70—90		" "	7.66	+633	» »	0.32	0
160—180		" "	7.56	+579	» »	0.33	0
250—270		" "	7.62	+582	» »	0.36	0
4885	3973						
13—30		Mud, light brown	7.48	+571	—	—	—
53—75		Mud, chocolate color	7.54	+565	—	—	—
120—140		Mud, brown	7.56	+558	—	—	—
Upper 4886 part of dredge sample	4175	Mud, brown, effervesces in HCl	7.54	+572	—	—	—
Lower part		The same	7.48	+573	—	—	—
30—50		Mud, brown, soft, with dark layers	7.48	+557	—	—	—
90—108		Mud, dark brown	7.53	+550	—	—	—
195—216		The same	7.49	+572	—	—	—
4889	3424						
10—35		Mud, beige	7.66	+533	—	—	—
130—150		The same	7.68	+517	—	—	—
250—270		" "	7.77	+537	—	—	—

TABLE 1 continued

Station No. and layer in core and dredge sample	Depth, m	Description of sediment	pH	Eh, mV	Fe··	Fe···	Fe··/Fe···
4896	3785						
Middle layer of dredge sample		Mud, straw color, with fragments of volcanic rocks	7.54	+554	—	—	—
4897	4244						
Middle layer of dredge sample		Mud, dark brown, soft, many corrections	7.60	+594	—	—	—
20—40		Mud, brown	7.45	+546	—	—	—
80—101		The same	7.45	+574	—	—	—
157—176		" "	7.41	+576	—	—	—
4901	5378						
Upper part of dredge sample		Mud, argillaceous, dark brown	7.21	+571	Not det	0.62	0
Lower part		The same	7.19	+582	" "		
30—57		" "	7.13	+557	" "	0.79	0
120—140		" "	6.91	+586	" "	0.57	0
212—232		" "	6.91	+604	" "	0.59	0
268—290		" "	6.99	+623	" "	0.56	0
4905	5151						
Middle layer of dredge sample		Mud, brown	7.40	+522	" "	0.45	0
68—88		Mud, light brown	7.29	+435	" "	0.43	0
130—150		The same	7.25	+429	" "		
180—203		" "	7.26	+461	" "	0.48	0
285—310			7.26	+454			
397—421		Mud, brownish gray	7.25	+467	" "	0.34	0
574—600		The same	7.25	+454	" "	0.40	0
715—742		" "	7.20	+520			
825—851		The same, with black spots	7.22	+504	" "	0.41	0
935—962		The same	7.26	+468			
1035—1060		The same, with greyish spots	7.23	+508	" "	0.37	0
1129—1151		Mud, light brown	7.29	+490	" "	0.38	0
5003	5685						
7—17		Mud, silty-argillaceous, radiolarians, brown	—	+494	0.09	0.46	0.20
90—113		Mud, silty-argillaceous, dark gray	—	—346	0.51	0.19	2.68
145—166		The same	—	—343			
230—250		The same, gray	—	—344	0.81	0.14	5.78
276—300		The same	—	+25			
334—350		Mud, silty-argillaceous, brown, with dark brown spots	—	+594	No det	0.64	0
360—383		Mud, silty-argillaceous, gray	—	—139	0.39	0.31	1.26
472—495		The same, greenish gray	—	—100	0.83	0.15	5.53
560—582		The same	—	—259			
595—618		The same, gray	—	—9	0.63	0.10	6.30
687—710		The same	—	—330			
760—786		The same, dark gray	—	—334	0.64	0.08	8.00
5005	4569						
285—53		Mud, straw color, with rose tone, effervesces in HCl	7.30	+537	—	—	—
80—104		Mud, brown	7.22	+537	—	—	—
130—155		The same	7.22	+552	—	—	—
		Bay of Bengal					
Upper 4925	3482						
part of dredge sample		Mud, grayish brown	7.52	+436	No det	0.71	0
Lower part		The same	7.66	+331	—	—	—
50—69		Mud, dark gray	7.45	—261	0.78	0.12	6.50
150—170		The same, with hydrotroilite, very weak odor of H_2S	7.38	—132	1.66	0.16	10.38
200—220		Mud, dark gray, thin layers	7.30	—127	—	—	—
300—320		Mud, dark gray, with mica	7.24	—140	0.87	0.10	10.04
425—446		Mud, dark gray, with hydrotroilite	7.20	—90	1.86	0.15	12.33
530—555		The same	7.20	—83	—	—	—
640—665		" "	7.24	—137	2.19	0.13	16.84

TABLE 1 continued

Station No. and layer in core and dredge sample	Depth, m	Description of sediment	pH	Eh, mV	Fe$^{\cdot\cdot}$	Fe$^{\cdot\cdot\cdot}$	Fe$^{\cdot\cdot}$/Fe$^{\cdot\cdot\cdot}$
4933	2029						
20—40		Mud, gray, with bluish tone	7.50	−358	—	—	—
70—91		The same	7.51	−335	—	—	—
190—210		Mud, gray	7.54	−359	—	—	—
300—320		The same	7.54	−266	—	—	—
4936	3103						
2—13		Mud, calcareous, silty-argillaceous, light brown	—	+442	0.04	0.48	0.08
43—60		Mud, gray, dense	—	−52	—	—	—
60—80		Mud, argillaceous, gray	—	−19	1.34	0.18	7.44
130—150		Mud, dark gray	—	−116	—	—	—
200—220		Mud, argillaceous, dark gray, with admixture of hydrotroilite, with weak odor of H$_2$S	—	−105	1.69	0.17	9.94
301—312		Mud, calcareous, silty-argillaceous, dark gray, with small inclusions of hydrotroilite, with odor of H$_2$S	—	−176	1.75	0.12	14.58
395—405		The same	..	−158	1.32	0.11	12.00
Lower 4969 part of dredge sample	2940	Mud, silty-argillaceous, gray	7.56	−349	—	—	—
30—42		The same	7.58	−363	1.09	0.38	2.87
66—81			7.62	−367	—	—	—
110—135		The same, rare inclusions of hydrotroilite	7.65	−368	1.11	0.24	4.62
200—217		The same	7.69	−318	0.99	0.22	4.50
385—400		The same, thin layers of silt and hydrotroilite, weak odor of H$_2$S	7.64	−350	1.21	0.52	2.33
435—455		The same	7.67	−332	1.15	0.50	2.30

true for the basins on the Taman Peninsula, for example, in which the sediments, judging from sample descriptions by Romm [1950], are black and are consequently characterized everywhere by the development of hydrotroilite, and for which a great scattering of points is observed, as seen in Fig. 4. Here the points are grouped along the ordinate axis, which attests to an absence of any correlation between Fe$^{\cdot\cdot}$/Fe$^{\cdot\cdot\cdot}$ and Eh.

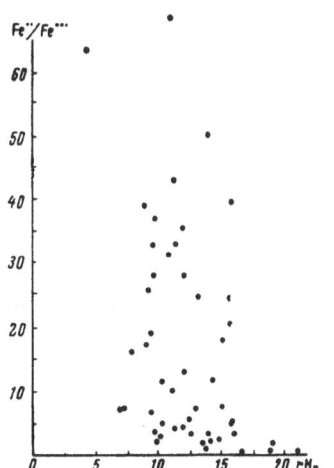

Fig. 4. Correlation between the iron ratio and the value of rH$_2$ in recent sediments of the Taman' estuary (plotted from data of Romm [1950]).

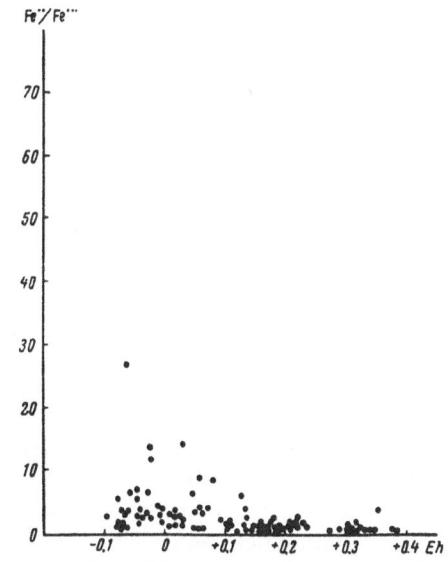

Fig. 5. Correlation between iron ratio and the oxidation−reduction potential in muds of the Sivash (plotted from the data of Stashchuk, Suprychev, and Khitraya [1964]).

For the sediments of the Sivash, where hydrotroilite is also everywhere present, the points are grouped chiefly along the abscissa axis, which indicates the absence of any correlation between the investigated values (Fig. 5). Unfortunately, we have found no data in the literature on recent subaqueous

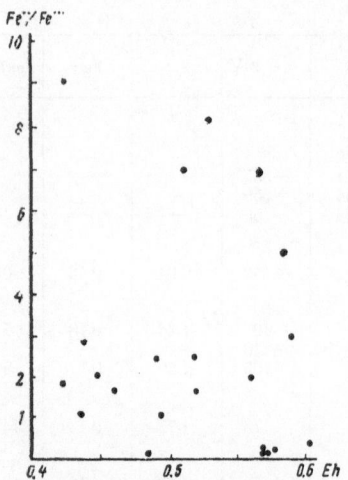

Fig. 6. Correlation between iron ratio and oxidation—reduction potential in soils from the Saratov Oblast (from the data of Poddubnyi [1959]).

sediments, which should reflect the field of development of iron hydroxides. Such a field might appear from recomputation of analyses of recent soils. Thus, use of data from Poddubnyi [1959] on soils from the Saratov Region has allowed us to plot Fig. 6, in which it is very obvious that there is no correlation between $Fe^{\cdot\cdot}/Fe^{\cdot\cdot\cdot}$ and Eh.

The entire complex of facts thus fails to point up any quantitative relationship between the iron ratio in the solid phase of the sediment and the oxidation—reduction potential. We may merely speak of a not very clear break between $Fe^{\cdot\cdot}/Fe^{\cdot\cdot\cdot}$ and Eh, existing between the field of iron hydroxides and the field of iron sulfides. This break cannot be defined quantitatively. As seen from Figs. 2-5, iron sulfides begin to form in different basins at different values of Eh. Therefore, the use of the ratio $Fe^{\cdot\cdot}/Fe^{\cdot\cdot\cdot}$ for determining oxidation-reduction conditions becomes qualitative, and is no better than the color characteristic of the sediment, which may be obtained with much less work. As a consequence, we are forced to the conclusion that there is no basis for accepting the method of determining oxidation—reduction conditions from the iron ratio in sedimentary rocks, since it has not proved valid even for different types of recent sediments.

It is no wonder, therefore, that analysis of extensive factual material collected by V. G. Savich and O. P. Chetverikova on the physicochemical characteristics of clays, limestones, sandstones, and siltstones of Paleozoic, Mesozoic, and Cenozoic ages from the Kuybyshev, Saratov, and Volgograd regions led Sokolov [1962] to the conclusion that

no relationship has been established between the oxidation—reduction potential and lithic type of rock, or presence of gas or oil, or the $Fe^{\cdot\cdot}/Fe^{\cdot\cdot\cdot}$ ratio. From an entirely different viewpoint, particularly from theoretical considerations, Kliburszky [1958] concluded that it is impossible to use the Fe_2O_3/FeO ratio obtained from analysis of solid phases to determine the oxidation—reduction potential under any conditions.

The Forms of Sulfur as Indicators of Oxidation — Reduction Conditions

Even less definitive are conclusions that the oxidation—reduction potential depends on the various forms of sulfur bound in the solid phases of sediments. In referring to Romm's data [1950], Strakhov wrote: "except for some particular variations, the generation of H_2S in sediments is directly related to the content of organic matter and rH_2; with increase in C_{org} and decline in rH_2 the ions of bound and free H_2S increase" [1954, p. 582].

However, later data of Volkov [1961a] on the Black Sea indicate that there is no direct relationship between the content of free H_2S and the total amount of organic matter in sediments. In regard to bound sulfur, this investigator concluded that the amount of sulfur in sulfide is no indication of the intensity of reducing processes [Volkov, 1961b].

Generally speaking, we should note that Volkov, who concerned himself for a long time with a study of the distribution of different forms of sulfur in the Black Sea sediments, was unable to work out any clear relationships. We may judge of this uncertainty by the fact that, finding no correlation between organic matter and free hydrogen sulfide, on the one hand, or between sulfur in sulfide and the oxidation—reduction potential, on the other, this investigator has generally held to the view that sulfur is the most sensitive indicator of the oxidation—reduction potential [Volkov, 1959]. Such uncertainty in solving the problem may be an indication primarily of that fact that data are not always unambiguous, and thus do not permit us to adopt firmly any definite position. One of the principal reasons for the uncertainty of past conclusions has been the weak basis for the analytical method in determining different forms of sulfur in recent sediments and sedimentary rocks.

The fact is that during decomposition of a sample by acid, as a result of change in P_{H_2S}, tervalent iron is reduced rather easily, converting hydrogen sulfide to elemental sulfur. Therefore, determination of the different forms of sulfur (also, incidentally, determination of the forms of iron) becomes uncertain. Ostroumov [1953a] has proposed a method of determining the different forms of sulfur from a single sample, a method that is applicable to recent bottom deposits. Although no limits are set in the paper for use of the method, it may be used only when tervalent iron is absent. It is clear that, from these viewpoints, the data furnished by Romm [1950] and Savich [1950] cannot serve as sufficiently reliable support for analyzing the systematic relationships here investigated.

The reliability of the data on the Black Sea is justified by Ostroumov [1953b] by the fact that readily soluble sulfides of ferric iron cannot exist in the presence of hydrogen sulfide there, a fact attested, in the author's opinion, by the much greater amount of ferrous iron in readily soluble form (extracted by dilute sulfuric acid) than in the form of FeS. Proof of this has also been advanced by Volkov [1961b], who shows that investigation by the Stokes' method gave a negative reaction to ferric sulfide.

In the section Mineralogical Forms of Iron Sulfides in Sedimentary Rocks and Recent Sediments (Chap. 3) we touch on these questions from a somewhat different viewpoint. Here we note simply that Treadwell [1948-49], who has analyzed the Stokes' reaction, gave proof only for Fe_2S_3, a rapidly decomposing sulfide with decrease in P_{H_2S}. This by no means signifies that an ammoniacal solution of zinc oxide will decompose all sulfides in which ferric iron is present (such as $Fe_3S_4 = FeS \cdot Fe_2S_3$, and so forth), even as it does not decompose ferrous sulfides. But, even with Fe_2S_3, the Stokes' method

does not always give reliable results [Kumai, 1958]. Ostroumov [1953a] has found a number of objections to the method itself among a number of investigators [Kaplan, 1963; Berner, 1964a]. Apart from everything else, the authors of the method by which the oxidation–reduction potential of sediments may be evaluated from the forms of sulfur might convince themselves of the unreliability of the analytical results of determining the forms of sulfur in the very same data [Volkov and Ostroumov, 1957; Volkov, 1961b].

The Relationship of Organic Matter in Sediments and the Oxidation – Reduction Potential

The relationship between content of organic matter in sediment and Eh is as weakly defined as the relationship between the forms of sulfur and Eh. Strakhov undertook to derive a general rule, reflecting the relationships among these values. In analyzing the data of Savich [1950], Strakhov [1954] concluded that, with any particular content of organic matter, Eh declines with decrease in average diameter of the clastic particles constituting the sediment. In sediments of a particular grain size, Eh exhibits inverse dependence on the organic content. Salinity of a basin complicates this relationship: "fresh-water lakes with sediments poor in organic matter exhibit the highest value of Eh in the sediments; sapropelic lakes are distinguished by lower values of Eh, and saline lakes and seas are characterized by still lower values" [Strakhov, 1954].

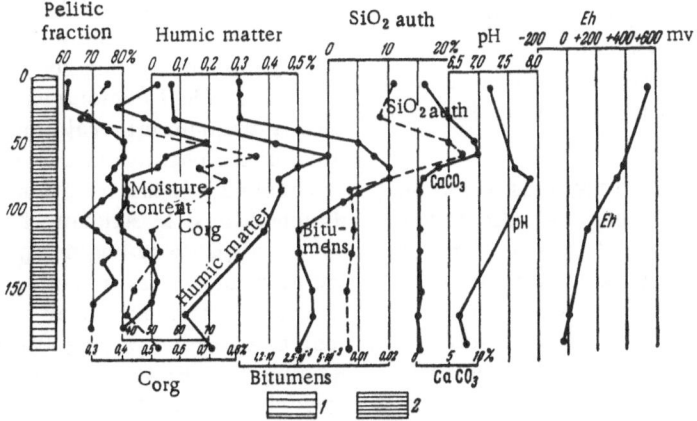

Fig. 7. Core from station 3325; Pacific Ocean, Obruchev Rise. Vertical distribution of the pelitic fraction, moisture content, C_{org}, and humic and bituminous substances. 1) silty-clay rocks; 2) clayey muds (after Romankevich [1960]).

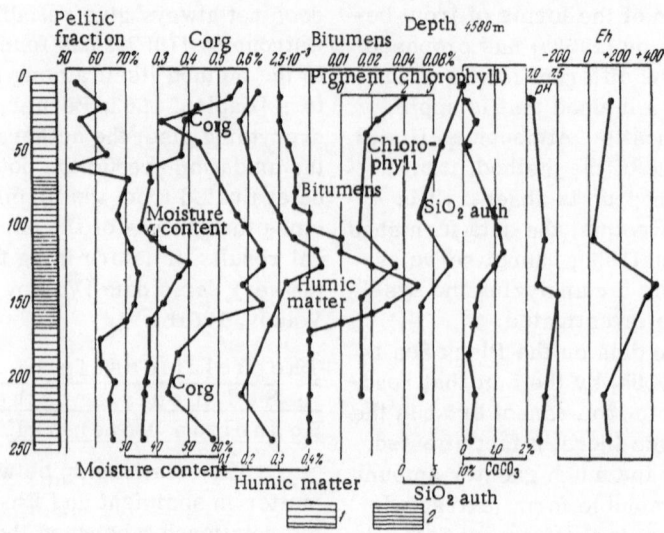

Fig. 8. Core from station 3342; Pacific Ocean, region of the junction of the Kurile—Kamchatka and Aleutian Trenches. Vertical distribution of pelitic fraction, moisture content, C_{org}, and humic and bituminous substances. 1) silty-clay muds; 2) clayey muds (after Romankevich [1960]).

However, these ideas have not explained all the complexities of the behavior of organic matter in the attempt to relate this to the Eh value. In particular, this may be seen from the data of Romankevich [1960], obtained from drilling into the muds of the Pacific Ocean (Figs. 7 and 8). As seen from Figs. 7 and 8, especially notable for station 3342, the oxidation—reduction potential does not always decline with increase in amount of organic matter and its various components, but may even increase. In one of his latest papers Strakhov finally concluded that "this relationship is, however, merely general and qualitative, not quantitative" [Strakhov, 1960b, p. 419]. This conclusion offers no possibility of using the content of organic matter in rocks (and sediments) as a standard of the oxidation—reduction processes taking place in the muds.

Thus, all the investigated methods based on comparison of individual elements as indicators of oxidation—reduction conditions during the formation of sedimentary rocks prove to be unsuitable for quantitative evaluation. Joint use of the various methods also proves to be relatively ineffective. An attempt to use several factors (the ratios of different forms of sulfur, iron, and organic matter) for determining the oxidation—reduction potential was made by Yurkevich, who later noted that "the expression of the ratio has proved to be rather complex, and the practical effect unsatisfactory. In this

connection, we should reject ratios of such type" [Yurkevich, 1958, p. 7].

Iron-Mineral Indicators of Oxidation — Reduction Conditions

The current view concerning minerals as indicators of oxidation—reduction conditions has been growing by accretion for a long time. But the greatest contribution to the actual character of the present viewpoint appeared at the beginning of the thirties in the work of L. V. Pustovalov. Even before publication of the well-known work of this investigator concerning geochemical facies, there were individual communications expressing the fact that authigenic minerals may serve as good indicators of the physicochemical conditions in the environment. In this relation, considerable credit belongs to A. D. Arkhangel'skii.

At the very beginning of the thirties, authigenic minerals as indicators of the oxidation—reduction conditions under which rocks form received ever-increasing attention and recognition by geologists. Criteria had already been noted for discrimination of general oxidation environments and general reducing environments. Sulin and Varov [1932] wrote that various chemical components might serve as indicators of a reducing character of the environment. Arkhangel'skii, in analyzing the conditions under which oil in the Northern Caucasus formed, turned his attention to indicators of the hydrogen-sulfide reducing conditions, appearing as finely comminuted pyrite and chemically precipitated calcium carbonate, drewite. The content of pyrite or other sulfur compounds will always be very high in rocks that formed in a hydrogen-sulfide environment. In contrast to the reducing complex, the complex of oxidizing conditions will be characterized by the presence of nitrogen oxides, ferric iron compounds, sulfates ($CaSO_4$), higher manganese oxides (MnO_2), and a small content of organic carbon, or with a large content of organic carbon in the form of coaly compounds. In 1933 the paper of Pustovalov [1933b] on geochemical facies was published. In this paper Pustovalov distinguished a whole series of minerals by which it was possible to characterize the oxidation−reduction conditions under which sediments formed. Let us recall briefly the characteristics of facies as proposed by Pustovalov.

1. Hydrogen-Sulfide Facies. This facies is recognized by the discovery of large amounts of pyrite in the rock. This is related to reduction of calcium sulfite and, consequently, to a sharply reducing environment. On the basis of the view that the oxidation−reduction potential declines gradually, starting at the surface of the water in the basin, this facies suggests that the oxidation−reduction boundary is found high above the sediments.

2. Siderite Facies. This facies is distinguished by the presence of siderite. The high content of CO_2 is associated with decomposition of organic material and, consequently, with anaerobic conditions, because of which the oxidation of iron is prevented. The oxidation−reduction boundary is found not far above the sediments.

3. Chamosite Facies. This is characterized by the iron silicate chamosite.

4. Glauconite Facies. This is characterized by minerals with varying contents of ferric and ferrous iron. Pustovalov believes that a definite ratio of ferric to ferrous forms of iron is produced in the mineral because of a constant struggle between oxidizing and reducing tendencies. As a consequence, it follows that the oxidation−reduction boundary lies within the sediments.

5. Oxidizing Facies. This facies is characterized by brown ferric oxide and manganese peroxide. There is an excess of oxygen, and the boundary of the oxidation−reduction potential consequently lies deep in the sediments.

The discrimination of geochemical facies by means of authigenic minerals is the result of long

years of studying and analyzing the conditions under which sedimentary rocks have formed. As one may learn from the work of Pustovalov [1933a], this investigator was greatly influenced in distinguishing geochemical facies by his views concerning the formation of minerals in recent sediments, information about which was rather scarce at that time. The work of Pustovalov on geochemical facies [1933b] has had great significance. In it, it was shown properly, for the first time, how it is possible to find certain factors, defining an assemblage of minerals, by means of which it is possible to investigate the physicochemical conditions under which sedimentary rocks formed.

Unfortunately, the discrimination of each facies has not been supported by convincing physicochemical considerations. Fersman [1955] and Pustovalov [1933b] therefore warned that the given list is preliminary and does not embrace all varieties of geochemical conditions. However, the logical completeness of the scheme has proved so attractive that it has been used widely with no substantial changes and is presently being used in practical geological work.

Pustovalov's views on geochemical facies became widely popularized in the works of Teodorovich [1946, 1947, 1956, 1964]. As an analysis of these works of Teodorovich shows, attention is given, first, to the existence of the boundary of oxidation−reduction conditions and the continuous decline in oxidation−reduction potential downward from this boundary, second, to detailed definitions of the geochemical facies themselves, grouping minerals relative to the form of iron, and, third, to the increased role of pH in the formation of mineral indicators.

Naturally, the views advanced by Teodorovich were directed at the objective of proving the possibility of distinguishing geochemical facies by mineral indicators. However, the principal conditions that are the primary objective of explanation and on which the proof of a change in minerals is essentially based were not discussed. In particular, no question was raised concerning a strict physicochemical basis for the change in minerals. It has been held, as an a priori view, that the reducing character of the medium must increase with depth. The sequence in mineral changes has been formulated only on this basis.

Considerable time has passed since the idea first appeared that mineral indicators might define oxidation−reduction conditions. Over this protracted interval, sufficient quantitative data have accumulated to attest to the fact that the sequence of formation of authigenic iron minerals as proposed by the scheme is not always encountered and does not always explain the facts of actual mineral paragenesis. In particular, it remains incomprehensible why a whole complex of minerals is frequently found in sedimentary rocks: pyrite, siderite, glauconite (and, commonly, magnetite), all together. Why is it that early formed minerals are not converted to new forms in keeping with the conditions imposed on them by the decline in potential? If this conversion does not take place, where does the material come from for formation of new minerals if such material must have been completely used in the minerals formed earlier?

Some investigators, in this connection, state that iron hydroxide brought into the basin is for the most part in a colloidal form relatively unsusceptible to reduction, and this form is preserved without change, depending on the intensity of the reducing processes during diagenesis of sediments (Kuz'mina, Maimin, Petrova). Others, on the other hand, consider iron hydroxide to have the highest reactive capacity, and for their genetic conclusions they use hydrochloric-acid extraction from the rock, by means of which these forms of iron are completely removed [Strakhov and Zalmanzon, 1955]. According to this idea, new minerals form as a result of the solution of previously formed minerals [Strakhov, 1956].

Lastly, a third group try to explain the coexistence of the minerals by the specific effect of processes taking place on the floor of the basin. In particular, Gulyaeva [1956] has stated that if CO_2 forms in an oxidizing environment relative to organic matter then siderite is formed. If the formation of CO_2 proceeds by decarboxylization, this process may take place in a reducing environment relative to organic matter, as a result of which pyrite will form instead of siderite. It is therefore concluded that pyrite and siderite may be paragenetically related and that, in this case, the oxidation—reduction boundary must be above the sediment all the time. Gulyaeva has shifted her opinion from mineral indicators to amount of the sulfide form of sulfur as an index to the reducing character of the environment.

In recent years we have begun to collect data indicating that the paragenesis of mineral indicators is not at all connected with an obligatory decline of the oxidation—reduction potential. For example, it has been noted that glauconite and pyrite are present together in recent marine sediments, both of them authigenic [Aleksina, 1962]. Babaev [1957b] noted the applicability of this feature to sedi-

mentary rocks, though it is true that he could find no sufficient basis for the phenomenon.

Larskaya [1961] has written concerning the existence of pyrite—glauconite and glauconite—pyrite geochemical facies. There is a noteworthy remark of Klud's that glauconite is found together with pyrite and that recent sediments containing glauconite give off a hydrogen-sulfide odor. Similar remarks were made by Gallier [1935], who studied modern processes of glauconite formation.

We may allude to the work of Bushinskii [1937], who studied the petrography of the Egorev phosphorites and drew the following conclusions. Glauconite separated out first (ferruginous oolites in the Ryazan horizon), then concretions of amorphous and microcrystalline phosphate appeared. It is possible that some of the pyrite formed at this time. Cavities in the phosphorite were then coated by crusts of radiating phosphate, and, finally, the remaining cavities in the phosphorite were partly filled with pyrite and pigmentary glauconite. In barren rocks pyrite began to separate at the same time granular glauconite formed. From this detailed description it appears that pyrite and glauconite might be syngenetic and not related to successive lowering of the potential, but the remaining minerals (siderite, magnetite, leptochlorite), despite the presence of CO_2 (phosphorite is formed), are excluded.

Strakhov approached the possibility of the formation of mineral indicators of oxidation—reduction conditions from a somewhat different position. Whereas in Pustovalov's scheme, developed subsequently by Teodorovich, the main cause of mineral change is the successive lowering of Eh with depth and the related progressive reconstitution of the minerals toward increase in the $Fe^{..}/Fe^{...}$ ratio, in Strakhov's scheme the principal criterion for change in minerals is the amount of organic matter (reducing agent) in the sediment and the assumption that the rates or difficulty of reactions differ according to the $Fe^{..}/Fe^{...}$ ratio in the mineral.

Strakhov has given a detailed description of this process [Strakhov and Zalmanzon, 1955]. The essence of the process is the following: the leptochlorite reaction not only begins earlier than others, but it requires a minimal expenditure of reducing agent (organic matter); it could therefore take place easily and quickly. The siderite reaction begins somewhat later and takes place with greater difficulty, at a slower rate, because it requires greater expenditures of the reducing agent. Pyrite not only begins to form after the others (according to Eh) but it forms much more slowly, with greater difficulty than all the others, because it requires the greatest expenditure of reducing agent.

The slowness of the processes when organic matter is present in sufficient quantities must determine the sequence of mineral formation in a manner similar to that proposed by Pustovalov. The hypothesis advanced by Strakhov makes it possible to explain the joint occurrence of several minerals. Actually, with low C_{org} content the reduction of ferric iron quickly ceases and the sediment will be dominated by leptochlorite, with small admixtures of siderite and negligible quantities of pyrite. With increase in organic matter, the series shifts toward greater amounts of pyrite.

From these statements it becomes clear that the method of determining oxidation–reduction conditions by the amount of pyrite or pyritic sulfur extracted from the rock during chemical analysis [Gulyaeva, 1956] may here be considered a particular case of the connection between organic matter and the rate of formation of minerals differing in the state of reduction of the iron. This variation of the method was subjected to some critical remarks by Strakhov [Strakhov and Zalmanzon, 1955], however, and was shown to be completely unsuitable for studying Lower Carboniferous rocks in the Kama and Middle Volga regions [Yagofarov and Gorelova, 1962].

Unfortunately the method proposed by Strakhov cannot give a reliable determination of the oxidation–reduction potential. As the author of the method himself noted, the parameter Eh in this case defines the lower boundary and may be considered an approximate semiquantitative value [Strakhov and Zalmanzon, 1955].

Garrels' Thermodynamic Diagrams

In 1952 Krumbein and Garrels [1952] published a paper in which there first appeared a diagram

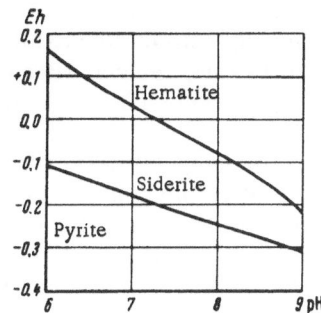

Fig. 9. Stability of iron minerals (after Krumbein and Garrels [1952]).

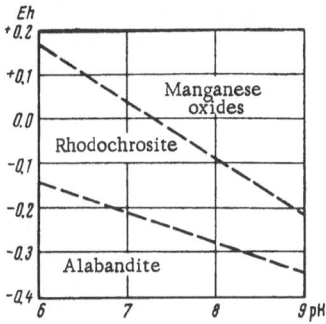

Fig. 10. Stability of manganese minerals (after Krumbein and Garrels [1952]).

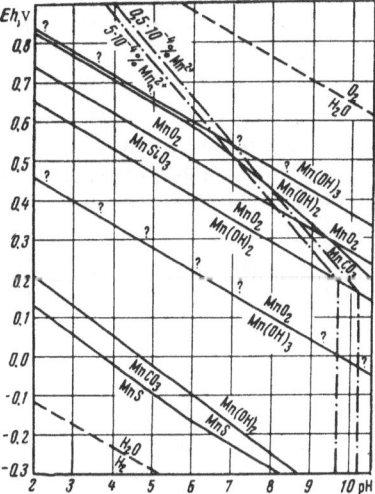

Fig. 11. Stability of manganese minerals (after Krauskopf [1957]).

based on thermodynamic calculations and designed to explain the sequence of formation of iron minerals (Fig. 9) and manganese minerals according to Eh and pH values (Fig. 10). A more detailed diagram for manganese minerals (Fig. 11) has been computed by Krauskopf [1957]. These diagrams have given firm support to the theory of mineral indicators.

The computational part proved to be so well founded that the diagrams were immediately accepted by geologists and have become widely circulated in the geologic literature.

A similar type of diagram began to appear in the works of Huber and Garrels [1953], Baas Becking and Moore [1961], Tischendorf and Ungethüm [1965], Strakhov [1953, 1960a, 1960b, 1964], Teodorovich [1956], Garrels [1965], and others. The appearance of diagrams on which fields of mineral occurrence were bounded by Eh—pH coordinates made it possible to compare calculated behavior with actually observed facts by Eh and pH measurements and by the control of authigenic minerals in recent sediments.

Measurements have shown that the theoretical diagram by no means corresponds to the conditions of mineral formation observed in the present-day growth of minerals. In 1963 and 1964 Semenovich published papers showing that the sulfide muds of Lake Ladoga do not fit in fields on the Garrels diagram. Study of the Black Sea muds has shown that sediments rich in hydrotroilite occur at the surface of the sea floor in those places where reducing processes are very weak and free sulfur is practically absent [Volkov, 1964].

A serious indirect objection to Garrels diagrams is the absence of siderite and alabandite in all recent marine sediments, although the diagram shows a field of siderite occurrence of appreciable size and with limits of Eh and pH frequently encountered in present-day sediments. It is generally thought that this discrepancy may be due to the approximate determination of Eh and pH values by Krumbein and Garrels [Pustovalov and Sokolova, 1957; Sokolova, 1962], incorrect preparation of the rock for determination of Eh, or incorrect choice of electrode [Solomin, 1964]. Garrels himself later wrote: "It would be unfortunate if attempts are made to read into these diagrams a validity and accuracy that are not there" [1960, p. 219].

The fact is, apparently, that not only the data used for plotting the diagrams are insufficiently precise, but the processes at work in the muds are more complex than those considered in plotting the diagrams. It is important to keep in mind that the results of mathematical and physicochemical treatment of facts (the final result of which are the diagrams), however the treatment is carried out, are determined by data that may not always be sufficiently reliable. The source of discrepancies between computed data and observed facts must therefore be sought primarily in those premises that form the basis of the computations. It is this position from which Garrels later began to consider the diagrams.

In 1963, for example, the author first came out with a criticism of the diagrams in connection with the fact that the reaction involved in oxidation of hydrogen sulfide to sulfate and used in plotting the diagrams is irreversible [Stashchuk, 1963]. Similar objections have been given by Zavodnov [1965]. A more thorough investigation of this problem has been discussed in the work of Stashchuk

[Stashchuk et al., 1964]. All this impels us to be careful in our reference to conclusions that may be drawn from Garrels thermodynamic diagrams.

Other Methods of Determining Oxidation — Reduction Conditions

The investigated methods may give, at best, a qualitative definition of oxidation—reduction processes taking place in muds at the time rocks are being formed. This means that not one of the above criteria can serve as a firm basis for quantitative comparison of different stratigraphic sequences or even of different horizons within a single sequence according to physicochemical characteristics. And it follows that we must search for new methods that may permit us to solve this problem. We may note three directions of research that have begun to develop especially in recent years:

1. direct electrometrical determination of oxidation — reduction potential in rocks;
2. determination of reduction capacity;
3. a composite method of determining oxidation — reduction conditions.

A detailed examination of electrometrical methods would require of the reader rather clear physicochemical views concerning oxidation—reduction potential and the components affecting it. Since information on this question is given in the following chapters, we consider it inadvisable here to consider the peculiar features of these methods. In our opinion, not one of the variants of existing electrometrical methods applicable to sedimentary rocks for the purpose of determining Eh of the primary sediment can be considered to give a unique solution at the present time.

Some investigators have decided in general to abstain from any attempt to find a quantitative characteristic of oxidation—reduction conditions and have directed their efforts toward development of measuring the reduction capacity of rocks. Adherents of this view furthermore do not insist that the reduction capacity reflects the oxidation—reduction potential [Yurkevich, 1958; Bardoshi and Bod, 1960]. This value characterizes the state of reduction of the rock.

In this concept there is incorporated a quantitative evaluation of the maximum work accomplished by the rock during the conversion of all its reduced components to the oxidized state. Therefore, we should not need to pause on this method, perhaps, if the concept of reduction capacity were always perceived as the maximum work of oxidation reactions.

If we start from a concept of this value as used in chemistry, its application is justified since it may exhibit a good criterion for understanding epigenetic processes taking place through the influence of groundwater. The method of reduction capacity permits us to investigate the possibility of changes in this water according to reduction possibilities of the rocks, to predict epigenetic zones and geochemical conditions in the redistribution of elements, and so forth. It is precisely from this point of view that Evseeva and Perel'man [1962], Evseeva and Fomina [1965], Perel'man [1965], and others have approached this parameter.

It is an error, however, to consider the oxidation−reduction capacity to define the oxidation−reduction potential. It is just this treatment of the reduction capacity that has slipped into the works, for example, of Epatko and Shnyukov [1962], Laput' [1962], and others. Such an understanding of this view, however, unavoidably leads to the opinion that there must be a strict relationship between oxidized and reduced forms of iron preserved in the solid phase and the oxidation−reduction potential; and this, as we have seen from the preceding examples, is by no means always found to be true in practice.

The great variety of equivalent methods would seem to lead one to conclude that only a multiple course may supply the characteristic of oxidation−reduction conditions during the formation of a rock. For example, Khalatin [1961] combined the methods of Teodorovich, Gulyaeva, and Strakhov and Zalmanzon in reconstructing the oxidation−reduction conditions during formation of Lower Carboniferous rocks in the Kama-Kinel' basin. Slavin [1961] recommended the combined methods of Strakhov and Gulyaeva. Vainbaum [1960a] stated that only all indices taken together can define the oxidation−reduction conditions: the general facies conditions, the composition of authigenic minerals, the nature and composition of organic matter, the relations among the forms or iron and sulfur, the reduction capacity, and others. Yurganov [1956a] suggested that an entire series of determinations be made to find the

characteristic of the oxidation−reduction conditions, without explaining how to combine the results obtained or what they might actually mean in regard to the physicochemical peculiarities of the environment. Laput' [1962] tried to combine the oxidation−reduction potential and the reduction capacity.

It is precisely this trend, in which different individual methods, possibly contradictory, are combined for determination of the oxidation−reduction that points up with great eloquence the fact that there is still much that remains unclear in the problem. Many investigators have been forced to this very conclusion. For example, Yurkevich wrote that "Indicators of reducing conditions corresponding to the problems of petroleum geology and methods of studying such indicators have not yet been established. Some investigators believe that the state of reduction of a rock may be judged from its color and the richness of its organic matter, whereas others think that variation in the ferric−ferrous ratio of iron oxides in the rock or the content of reduced sulfur are more indicative. Attempts to express the state of reduction of a rock by means of measurements of the oxidation−reduction potential (Eh and rH_2) have gained no appreciable success" [Yurkevich, 1962, p. 128].

We believe that a sufficient amount of factual material has already accumulated, both for sedimentary rock and for recent sediment, that we may now form some systematic appraisal of our knowledge of this problem. We hope that, after such systematic evaluation, it will become possible to lay out a course for future work in determining the conditions under which sedimentary rocks form and to investigate the criteria that may serve as indicators of the oxidation−reduction conditions prevailing during formation and diagenesis of sediments.

In order to gain an understanding of the causes for the variance among the discussed criteria for evaluating oxidation−reduction potential, it is necessary, even if in the most general outlines, to acquaint ourselves with those physicochemical concepts and statements that we shall use in further discussion.

CHAPTER 2

THERMODYNAMIC PRINCIPLES FOR INVESTIGATING THE CONDITIONS OF MINERAL FORMATION

The Solubility of Compounds and the Maximum Work of a Reaction

Guldberg and Waage first studied the effect of the concentrations of components on the rate of a chemical reaction in 1867 and they formulated the rule that has received the name of the l a w of m a s s a c t i o n : the rate of chemical reaction is proportional to the "active masses" of the reacting substances.

For example, let us consider the equation in the general form:

$$A + B = C + D$$

In keeping with the above, the rate of reaction is proportional to the product of the reacting substances:

$$C_A \cdot C_B \cdot K_1 = V_1$$

where K_1 is the proportionality factor and C is the concentration, expressed in the components of the elements of the reaction A and B.

Let us complicate this example somewhat. Let us assume that the following reaction takes place:

$$2A + 3B = N + M$$

This reaction may be represented in the form

$$A + A + B + B + B = N + M$$

and, according to the law of mass action, we write

$$C_A \cdot C_A \cdot C_B \cdot C_B \cdot C_B = C_A^2 \cdot C_B^3 \cdot K = V$$

Thus, when components in a ratio not equal to 1:1 participate in a reaction, the rate of the reaction is determined by the product of the reacting substances in powers corresponding to their stoichiometric coefficients.

Let us return to the initial reaction:

$$A + B = C + D$$

At the very first moment of reaction the amount of final products C and D will be 0, and the product of the initial reaction products will be appreciably large. In keeping with this, the rate of the reaction toward formation of C and D will be appreciably large. With expenditure of substances A and B and the appearance of the reaction products C and D, reaction between the latter two becomes increasingly possible, shifting the reaction in the reverse direction. The rate of the reverse process is expressed by the equation

$$C_C \cdot C_D \cdot K_2 = V_2$$

where K_2 is a constant characteristic of the given substances.

During the reaction, finally, there comes a moment when $V_1 = V_2$ and, consequently,

$$C_A \cdot C_B \cdot K_1 = C_C \cdot C_D \cdot K_2$$

In this case, we say that stable dynamic equilibrium has been achieved. We may then write

$$\frac{C_C \cdot C_D}{C_A \cdot C_B} = \frac{K_1}{K_2} = K$$

This indicates that, under conditions of dynamic equilibrium, the product of the concentrations of substances that have formed and the concentrations of the initial substance is constant for a given temperature.

In a similar way we may examine the equilibrium conditions when the end products represent a solid phase. During the interaction between sediment and a solution, two processes take place: solution of the sediment and crystallization of dissolved substance. If we designate the solid salt by $K_a A_h$ and the solution products by K_a' and A_h', these two processes may be expressed by the equation

$$K_a A_h \rightleftarrows K_a' + A_h'$$

solid phase   solution

We shall designate by n_1 the number of cations and by n_2 the number of anions striking a unit surface of the solid phase, and we shall designate the number of cations and anions in solution by K'_a and A'_h respectively. The rate V_1 at which the cations will go into solution is proportional to this surface; i.e.,

$$V_1 = K_1 \cdot n_1$$

where K_1 is the proportionality factor.

At this same time the reverse process takes place: the precipitation of cations on the surface of the solid phase. The greater the number of anions per unit surface of the solid phase, the greater the number of cations that will precipitate. On the other hand, the higher the concentration of cations in solution, the greater the number also that will be precipitated. The rate of precipitation is thus directly proportional to the number of cations (concentration) in solution (K'_a) and the number of anions per unit surface area of the solid phase:

$$V_2 = K_2 \cdot n_2 \cdot K'_a$$

where K_2 is the proportionality factor.

When the rate at which cations go into solution is equal to the rate of their precipitation on the solid phase, i.e., when $V_1 = V_2$,

$$K_1 n_1 = K_2 \cdot n_2 K'_a$$

By a similar line of reasoning relative to the anions, it becomes clear that

$$K_3 \cdot n_2 = K_4 \cdot n_1 \cdot A'_h$$

where K_3 and K_4 are the proportionality factors respectively for anions going into solution and for their precipitation per unit area of solid-phase surface.

After multiplying the left and right sides of the equations thus obtained, we find

$$K_1 \cdot n_1 \cdot K_3 \cdot n_2 = K_2 \cdot n_2 \cdot K'_a \cdot K_4 \cdot n_1 \cdot A'_h$$

whence

$$K'_a \cdot A'_h = \frac{K_1 \cdot K_3}{K_2 \cdot K_4} = \text{SP}.$$

The solubility product (SP) is a characteristic constant for each compound at a fixed temperature.

Let us compare the value obtained for equilibrium constant and solubility product

$$\frac{C_C \cdot C_D}{C_A \cdot C_B} = K$$

If A + B represents solid material, then

$$K = \frac{C_C \cdot C_D}{AB} = C_C \cdot C_D = \text{SP}$$

In other words, when a solid phase forms, its concentration proves to be unity. It is necessary to take this into account in further considerations when we shall be speaking of more complex reactions.

By making use of the above considerations, let us analyze the dissociation of water. Observed dissociation of water takes place according to the scheme:

$$H_2O \rightarrow H^{\cdot} + OH'$$

The process in the direction of the arrow takes place only to a small degree. In agreement with the law of mass action, the rate of decomposition into ions is proportional to the active mass

$$V_1 = K_1 \cdot H_2O$$

but the rate of the reverse formation of water from $H^{\cdot}$ and OH' ions is proportional to the product of the active masses

$$V_2 = K_2 \cdot H^{\cdot} \cdot OH'$$

Under conditions of equilibrium $V_1 = V_2$, and therefore

$$K_1 \cdot H_2O = K_2 \cdot H^{\cdot} \cdot OH'$$

$$H^{\cdot} \cdot OH'/H_2O = K_1/K_2 = K$$

Since the decomposition of water into ions is very insignificant, the concentration of water may be considered a constant. Therefore, in its final form, the equation will appear in the following form:

$$H^{\cdot} \cdot OH' = K_{H_2O}$$

The most widely used methods in chemistry express concentrations in moles and gram-ions referred to a liter of solution or to kilograms of solvent. Let us recall that a mole, or gram-molecule, is the amount of substance equal to the molecular weight and a gram-ion is the amount of substance according to the weight of an equal sum of atomic weights of the elements constituting the given ion. The value of K_{H_2O} measured by different methods proves to be 10^{-14} (at 22°C). This means that in a liter of pure water at 22°C the product of the $H^{\cdot}$ ions and OH' ions constitutes 10^{-14} g-ions. During dissociation in pure water, the appearance of the $H^{\cdot}$ ion is necessarily accompanied by the appearance of the OH' ion. Therefore, for this water $H^{\cdot} = OH'$. Hence, it is easy to see that

$$H^{\cdot} = OH' = \sqrt{10^{-14}} = 10^{-7}$$

In other words, during the dissociation of pure water, there will be 10^{-7} g-ions of $H^{\cdot}$ (or about 10^{-7}g) per liter and 10^{-7} g-ions of OH' (or 1.7 ·

10^{-6}g). The $H^{\cdot}$ ion content in solution determines the acidity. The medium is considered neutral if $H^{\cdot} = OH' = 10^{-7}$; it is acid if $H^{\cdot} > 10^{-7}$. In chemistry the negative logarithm has been designated by the letter p. The negative logarithm of the dissociation (or equilibrium) constant is thus designated pK; the negative logarithm of the solubility constant is designated pSP, and the negative logarithm of the hydrogen-ion concentration is pH. Keeping these designations in mind, a neutral environment may be written as pH = 7, an acid environment as pH < 7. Since $H^{\cdot} \cdot OH' = 10^{-14}$, pH + pOH = 14 and pH = 14-pOH.

During solution of any substance, a definite interaction occurs between the ions in solution. Positive ions group around negative ions, and each negative ion is surrounded by a group of positive ions; i.e., the ions form an "ionic atmosphere" of oppositely charged ions about themselves. This ionic atmosphere, different each time according to the dilution of the solution and the ionic composition, gives the impression of incomplete dissociation of of the substance into ions during solution. When we speak of the equilibrium constant or the solubility product, our general concept is that ions of the substance that have passed into solution are completely free from the effect of bonds; i.e., we have in mind the conditions of ideal solutions. In considering actual solutions, it is necessary in each case to introduce some correction for the weakened capacity of ions for chemical reaction because of the retarding effect of electrostatic reactions. The correction must naturally be less than unity, approaching unity in dilute solutions. The corrected concentration, taking into account some retardation of the processes, is called the active concentration of a given ion or is called simply the activity, and is designated a. The activity of a given ion is thus proportional to its concentration and is always equal to or less than it.

$$a = \gamma \cdot C$$

where γ is the activity coefficient (molal scale) and C is the concentration in moles per liter.

Whence, if the reactions A + B = C + D or $A_2B_3 = 2A + 3B$ take place, we may write

$$\frac{a_C \cdot a_D}{a_A \cdot a_B} = K$$

in the first case and

$$a_A^2 \cdot a_B^3 = SP$$

in the second. Or

$$\frac{C_C \cdot C_D}{C_A \cdot C_B} \cdot \frac{\gamma_A \cdot \gamma_B}{\gamma_C \cdot \gamma_D} = K$$

$$C_A^2 \cdot C_B^3 = \frac{SP}{\gamma_A^2 \cdot \gamma_B^3}$$

In this book we shall almost always use values of ion activities.

Let us consider an example of using SP for calculating the possibility of ferric hydroxide going into ionic form. $Fe(OH)_3$ has SP = $4.5 \cdot 10^{-40}$; i.e., $a_{Fe^{\cdots}} \cdot a_{(OH)}^3 = 4.5 \cdot 10^{-40}$, or

$$a_{Fe^{\cdots}} = \frac{4.5 \cdot 10^{-40}}{a_{(OH)'}^3} \qquad (1)$$

In turn, taking into account the dissociation of water, we may write

$$a_{H^{\cdot}} \cdot a_{(OH)'} = 10^{-14}, \quad a_{(OH)'} = \frac{10^{-14}}{a_{H^{\cdot}}}$$

Substituting the value $a_{(OH)'}$ in Eq. (1), we find $a_{Fe^{\cdots}} = 4.5 \cdot 10^2 \cdot a_H^3$ or, in the logarithmic form, $\lg a_{Fe^{\cdots}} = 2.65 + 3\lg a_{H^{\cdot}}$ and, finally,

$$\lg a_{Fe^{\cdots}} = 2.65 - 3pH.$$

The activity of the ferric oxide ion in solution is thus controlled only by the pH value. By assigning arbitrary pH values we may compile a table of activities of the ferric ion in solution (Table 2).

As may be seen, the solubility of iron hydroxide at a pH of 1 is practically unlimited. The computed solubility of the ferric ion in such a solution is impossible, since the activity of the ferric ion is here controlled by other anions constituting the acid residue from dissolution (for example, $Fe_2(SO_4)_3$, where $a_{Fe^{\cdots}} = \sqrt{SP/a_{SO_4^{\cdot}}^3}$). With increase in pH values of one per unit, the solubility of iron hydroxide declines a thousandfold, and at a pH of 3 the activity of the ferric ion is measured in hundred thousandths of a gram per liter of water. At such constants, approximating natural waters, the activity may be equated with the concentration. Since such contents are not detected by ordinary chemical methods, we may then say, chemically, that at a pH of 3 complete

TABLE 2. Activity of the Ferric Ion According to pH Values

pH	$\lg a_{Fe^{\cdots}}$	$a_{Fe^{\cdots}}$ in g-ions/liter	$a_{Fe^{\cdots}}$ in g/liter
1	—0.35	$4.47 \cdot 10^{-1}$	$2.50 \cdot 10^{+1}$
2	—3.35	$4.47 \cdot 10^{-4}$	$2.50 \cdot 10^{-2}$
3	—6.35	$4.47 \cdot 10^{-7}$	$2.50 \cdot 10^{-5}$
4	—9.35	$4.47 \cdot 10^{-10}$	$2.50 \cdot 10^{-8}$
5	—12.35	$4.47 \cdot 10^{-13}$	$2.50 \cdot 10^{-11}$
6	—15.35	$4.47 \cdot 10^{-16}$	$2.50 \cdot 10^{-14}$
7	—18.35	$4.47 \cdot 10^{-19}$	$2.50 \cdot 10^{-17}$
8	—21.35	$4.47 \cdot 10^{-22}$	$2.50 \cdot 10^{-20}$
9	—24.35	$4.47 \cdot 10^{-25}$	$2.50 \cdot 10^{-23}$
10	—27.35	$4.47 \cdot 10^{-28}$	$2.50 \cdot 10^{-26}$

precipitation of iron hydroxide begins. Further increase in pH gives no perceptible increase in precipitate.

Geologists must be very careful in referring to statements involving chemistry. As we may convince ourselves from the cited example, the completeness of precipitation in the analytical sense still does not indicate complete elimination of the ferric iron from the solution.

Sometimes a very small amount of this ion is necessary to start a new compound, leading to complete disappearance of iron hydroxide. In particular, for Fe_2S_3 at a pH of 7 and $a_{H_2S} = 10^{-10}$ mole/liter (a value that cannot be detected by ordinary analytical methods), the activity of the ferric ion should be $1.7 \cdot 10^{-21}$ mole/liter. This value for the activity of the ferric ion is smaller by a factor of 250 than the value that guarantees iron hydroxide at the given pH. Therefore, in relation to Fe_2S_3, a solution containing iron hydroxide in the precipitate proves to be supersaturated, as a result of which Fe_2S_3 begins to settle out in the precipitate. But this leads to undersaturation of the solution relative to iron hydroxide. Iron hydroxide then begins to dissolve, and Fe_2S_3 again comes out in the solid phase.

We shall consider this example in more detail later. Here we simply wish to point out that those compounds that are considered insoluble by analytical chemists under the given conditions have a sufficiently large number of ions in solution under these conditions so that in the geologic sense they participate in migration and the formation of new compounds. From the geological point of view, therefore, we cannot neglect these low concentration values.

We are frequently confronted with the idea of maximum useful work of a reaction; let us therefore evaluate it. A compound that has formed is maintained by the force of reaction of atoms, groups of atoms, or ions. There is an oppositely directed force, due to thermal movement of particles, which tends to destroy this bond, to draw the particles away from the compound. The value that takes into account the resultant, or equilibrium, energy at constant pressure is called enthalpy. It is not possible to determine the enthalpy of a compound accurately. We therefore use values of enthalpy increments in thermodynamics. In this connection it is assumed that the enthalpy of elements in standard states is zero, where, by standard state, we mean conditions when the temperature is 20 or 25°C and the pressure is 1 atm. All remaining energy states are but varieties of enthalpy; i.e., each

is a comparative energy value of the unstable state. This value is called the enthalpy increment and is designated ΔH.

For example, molecular hydrogen is characterized by an enthalpy increment of $\Delta H = 0$. Atomic hydrogen has $\Delta H = 52,089$ cal. Consequently, with the transition $2H \rightarrow H_2$ the equilibrium energy will diminish by $0 - (2 \cdot 52,089) = -104,078$ cal $= \Delta H$; i.e., in order for atomic hydrogen to be maintained in the atomic state, specifically, in order for oscillatory movements to destroy the bond, it must have an energy 104,078 cal greater than molecular hydrogen.

As we have agreed, the value of enthalpy characterizes the equilibrium energy between the binding energy and the energy of disruption. Depending on the kind of particles, the state and nature of the bond, the mass of the particles, and other factors, the energy of thermal movement for different particles, tending to disrupt them, at any particular temperature may vary. In particular, for atomic hydrogen it may have one value, for molecular hydrogen another. Therefore, for computing the energy of the transition of molecular hydrogen to the atomic form, apart from the computed energy, it is still necessary to add a correction for the energy change of oscillatory movements in connection with the change in nature of the particles. This correction represents a measure of the disorder of the particles; it is called the entropy increment and is designated ΔS.

For atomic hydrogen the entropy at absolute zero and in the crystalline state is equal to zero. With each degree it increases 27.393 cal/mole. For molecular hydrogen, with each degree the value increases 31.211 cal/mole. Thus, the increment of disorder of a single mole of molecular hydrogen as compared with two moles of atomic hydrogen is expressed by

$$\Delta S = 31.211 - 54.786 = -23.575 \quad cal/deg\text{-}mole$$

The value obtained shows that, during the transition from atomic to molecular hydrogen, 23,575 cal more energy per degree of temperature is necessary in order to maintain the same intensity of thermal movement. Consequently, when atomic and molecular hydrogen are found at the same temperature, the bonds of hydrogen atoms in molecules require less energy than that computed for dissociation of atoms.

If the reaction takes place at 20°C, then

$$T = 273.16 + 20 = 293.16$$

where T is absolute temperature.

At this temperature the disruptive energy declines by $T\Delta S = 293.16 \, (-23.575) = -6911.247$ cal. Subtracting the correction we obtained, we determine the excess energy given off as a result of the transition of two moles of atomic hydrogen to one mole of molecular hydrogen. This energy may be used for performing useful work. Gibbs called it free energy. The value of the free energy is indicated by the letter G, and the change in free energy by ΔG.

Recently, according to a proposal of the Academy of Sciences of the USSR, a new terminology has become widespread. In this terminology the function G is called the isobaric–isothermal (or simply isobaric) potential. The term is related to the thermodynamic consequence of the function. In our opinion, a strict derivation is not necessary in this work. We have cited only one form of the logical explanation necessary for an understanding of the essence of G. In this connection, from the viewpoint of the indicated explanation, the concept of free energy is more suitable, since it characterizes the energy given off during a reaction at constant pressure, energy that may be used for performing work. In its general form ΔG represents not only the energy freed during the reaction but also, in general, all energy that may be used for obtaining work.

Thus, this course of reasoning indicates that

$$\Delta G = \Delta H - T\Delta S = 104{,}078 - (-6911.247) = 97{,}167 \text{ cal}$$

As a result of decline in resulting energy—enthalpy, and also as a result of the fact that the measure of disorder because of change in the character of the particles themselves (transition from atomic to molecular hydrogen) also diminishes, a large excess of energy is given off: 97.167 kcal. The excess of energy given off might be used partly or completely for performing useful mechanical, electrical, or other work; in this connection the value ΔG defines the maximum useful work of the reaction.

According to the first law of thermodynamics, energy can be neither created nor destroyed. It may be converted from one kind to another. If we ascertain that ΔG of any reaction is positive, it is then necessary to expend some amount of energy to effect this reaction; energy must be supplied from outside. If ΔG is negative, the reaction takes place with elimination of energy. No external source of energy is necessary for producing this reaction, and the reaction occurs spontaneously.

It must now be clear how great a role the concept of ΔG plays in physical chemistry.

For the reaction $2H \rightarrow H_2$ it may be said that at 20°C it will take place spontaneously, since no supply of external energy is necessary for it to occur. On the other hand, as a result of the occurrence of this reaction, excess energy will be given off, and this will preserve hydrogen in the atomic state.

Values of ΔG are given in chemical handbooks for many compounds.

Let us examine the possibility of forming water from molecules of hydrogen and oxygen:

	H_2	O_2	H_2O gas	H_2O liquid
ΔG^0 kcal	0	0	−54.635	−56.69

For the reaction to water vapor, the resulting free energy changes by

$$2\Delta G^0_{H_2O_{gas}} - (2\Delta G^0_{H_2} + \Delta G^0_{O_2}) = -109.27 \text{ kcal}$$

The investigated reaction must thus move toward formation of water. However, should this reaction take place under normal conditions, the air of the atmosphere would have been deprived of its hydrogen long ago, and a great part of its oxygen. Actually this does not take place because, for the reaction to take place, it is necessary first to expend work on converting molecules of hydrogen and oxygen to the atomic state, and then a reaction of the following type would take place:

$$2H + O = H_2O.$$

It has already been stated that, for the conversion of one mole of molecular hydrogen to atomic according to the scheme $H_2 \rightarrow 2H$, 97.167 kcal of energy must be expended. In order to split the oxygen molecule to give atomic oxygen according to the scheme $\frac{1}{2}O_2 \rightarrow O$, 54.994 kcal of energy must be expended. That is, the total energy expended for converting molecular hydrogen and oxygen to the atomic form is necessary in order to form one mole of water (vapor), and this is 152.161 kcal. During the formation of one mole of water, the resulting energy will be the difference between the energy expended on the conversion of molecular hydrogen and oxygen to the atomic form and the energy given off during the formation of the water:

$$-54.635 - (+152.161) = -206.796 \text{ kcal}$$

Taking into account the expenditure of energy on splitting molecular bonds, the energy given off will be 54.635 kcal per mole of water. This expen-

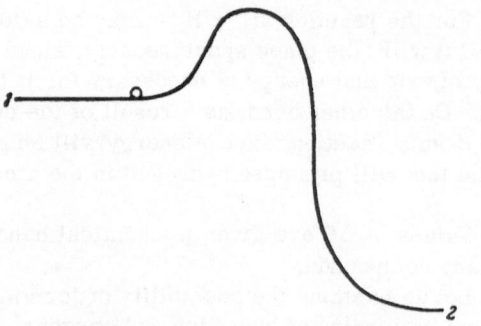

Fig. 12. Change in potential energy.

diture of energy is more than matched by the excess. But, as a matter of fact, at ordinary temperatures colliding molecules do not have sufficient energy to reach the threshold at which collision might rupture the molecular bond in sufficient numbers to permit this reaction to occur. Therefore, despite the appreciable gain in energy during formation of water from molecules of hydrogen and oxygen, the reaction does not take place. This process may be compared with the movement of a sphere along the surface illustrated in Fig. 12. The work of raising the sphere from level 1 up the rise will be compensated for by the excess when it rolls down to level 2. But the work of raising the sphere will be compensated at the time the sphere is rolling down the slope to level 2. The compensation does not exclude the fact that it is first necessary to raise the sphere and thus to expend a definite amount of work in raising it.

We have considered the value $-\Delta G$ as the energy freed during formation of a compound. The very liberation of energy indicates that the bond in the newly formed compound is stronger. Thus, the value $-\Delta G$ may also indicate the strength of the bond under given conditions. Let us consider another example: we may assume the following reaction in muds [Pel'sh, 1937]:

$$C_6H_{12}O_6 + 3H_2SO_4 \longrightarrow 6CO_2 + 3H_2S + 6H_2O$$
$$\Delta G = -180.96 \text{ kcal}$$

From the value of the standard free energy of the reaction, it is clear that the bond in the final products is stronger than the original bond, since the resulting free energy is negative. As a result of this, a considerable amount of energy is released, and the reaction may be formally considered to take place spontaneously. In practice, however, this reaction does not take place spontaneously under practical conditions, since, for it to occur, a definite amount of energy must be expended to break the bonds of the glucose and the sulfate ion. The

reaction is possible when sulfate-reducing microorganisms are active in the medium. In expending a definite amount of energy on breaking the bonds of the initial products, they receive a much larger amount of energy, taking care of the expenditure on breaking the bonds, and the excess energy obtained is used in the life activities of the organisms.

In summarizing what we have said, we should note the following: to explain the possibility of a reaction taking place it is necessary to write the reaction down and find the ΔG for each participating component from some handbook. From the sum of free energies of the final products, the sum of the free energies of the initial products is subtracted. Negative values of ΔG indicate that the reaction may take place in the direction in question. Positive values of ΔG indicate that the reaction may take place spontaneously in the opposite direction.

There is a logarithmic relationship between the investigated equilibrium constant (or SP) and the change in free energy. It has been shown in thermodynamics that $-\Delta G = RT \ln K$, where R is the gas constant and T is absolute temperature. Derivation of this equation becomes rather simple when we base it on the work performed by an ideal gas during expansion. Before going on to the proof, we should recall the following two assumptions upon which the proof is erected:

1. The Equation of an Ideal Gas According to Mendeleev. For one mole of ideal gas the equation is written in the form

$$P \cdot V = R \cdot T \qquad (2)$$

where P and V are pressure and volume of the gas respectively, T is absolute temperature, and R is a proportionality factor independent of temperature, type of gas, and pressure. The value R is called the universal gas constant. The above equation reflects the law that states that the product of the gas volume and the pressure on the gas is proportional to the temperature of the gas.

2. Natural Logarithms. In mathematics it is very convenient to use, not log to the base ten, but natural logarithms, which are indicated by the symbol ln. Common logarithms are to the base 10. For example, $\lg x = 2$ may be written as $10^2 = x$, $\lg y = 3$ as $10^3 = y$, and so on. In natural logarithms the base is $e = 2.7182818\ldots$. In this case, $\ln x = 2$ may be written as $e^2 = x$, $\ln y = 3$ as $e^3 = y$. Then, $\ln e = 1$.

The number e is computed from the function $(1 + 1/n)^n$. When $n \to \infty$, the function $(1 + 1/n)^n \to e = 2.7182818\ldots$. This may be readily verified by substituting different numbers for n.

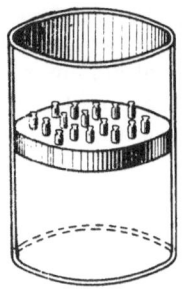

Fig. 13. Illustration for derivation of the equation for isothermal work of a gas.

Let us now go directly on to the proof. Let us make clear at the first step that the computation of maximum work of a gas during its expansion is made for constant temperature during the expansion. Figure 13 illustrates a cylinder with a piston that moves without friction, the cylinder containing an ideal gas of volume V_1. The gas in the cylinder is under a certain pressure P, which is produced by the weight of a large number of small weights placed on the piston. By removing the weights successively one after another, we make it possible for the gas to expand. When all the weights are removed, the volume of gas increases to V_2 and work is performed, the amount of which may be calculated as $A = f(P, V)$. The force used in moving the piston is determined as PS, where S is the area of the piston. The distance through which the piston may move is designated by Δh. Since the work is measured by the product of the force by the path distance, in the present case $A = PS\Delta h$.

But $S\Delta h$ is the change in volume of gas; i.e., $\Delta V = V_2 - V_1$. Should the external pressure remain constant while work is performed by the gas in the cylinder, then $A = P(V_2 - V_1)$. In the case in question, however, the pressure diminishes progressively with decrease in the number of weights. If the experiment is carried out at constant temperature, then, in keeping with Eq. (2), the pressure may be expressed as $P = RT/V$, and the work performed may be written in the form of the equation

$$A = RT \frac{V_2 - V_1}{V}$$

where V_2 is the final volume, V_1 the initial volume, and V the volume, variable during the performance of work by the gas, corresponding to the change in pressure.

Without knowing the nature of the change in this last value, having at our disposal only the final and initial volume, we cannot compute the total work performed by the gas. In order to escape this difficult position, we suggest that each weight placed on the piston have an infinitesimally small weight and

that the number of such weights be infinitely large. Then the removal of each weight produces an increase in volume of infinitesimally small value. As we have already pointed out, the performance of useful work by the system is characterized by a decrease in free energy of the system. Thus, if the initial state of the gas is characterized by a free energy of G_1 and a terminal free energy of G_2, the work performed should be defined by $G_2 - G_1 = A = -\Delta G$. For an infinitesimally small increase in volume, we may write

$$-dG = RT \frac{dV}{V_1} \tag{3}$$

Since dV has an infinitesimally small value, the right side of Eq. (3) may be multiplied, with no appreciable error, by

$$\ln \left(1 + \frac{1}{\frac{V_1}{dV}} \right)^{\frac{V_1}{dV}}$$

since, when $dV \to 0$, $V_1/dV \to \infty$ and

$$\ln \left(1 + \frac{1}{\frac{V_1}{dV}} \right)^{\frac{V_1}{dV}} \to \ln e = 1$$

Whence

$$-dG = RT \frac{dV}{V_1} \ln \left(1 + \frac{1}{\frac{V_1}{dV}} \right)^{\frac{V_.}{dV}}$$

After small transformations

$$-dG = RT \ln \left(\frac{V_1 + dV}{V_1} \right)$$

We now know definitely that V_1 is the volume corresponding to the energy state G_1 and dV is the increment of volume relative to V_1. In the given case there are no difficulties (in the sense of ambiguities in the solution) in dV increasing sufficiently that $V_1 + dV$ may take on the value V_2. But then $G_1 - dG = G_2$. In this case

$$G_2 - G_1 = -\Delta G = A = RT \ln \frac{V_2}{V_1} \tag{4}$$

Equation (4) makes it possible to compute the work performed by the gas during expansion only for initial and final volumes when temperature is constant. The work performed may be determined also by the change in gas pressure. Since

$$V_1 = \frac{RT}{P_1} \quad \text{and} \quad V_2 = \frac{RT}{P_2}, \quad \text{then} \quad A = RT \ln \frac{P_1}{P_2} \tag{5}$$

By using Eq. (5) we go on to the problem that directly concerns us, the proof of a connection be-

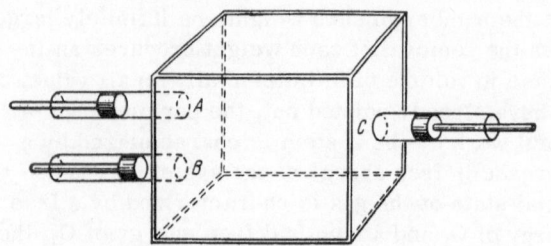

Fig. 14. The "Van't Hoff box."

tween the equilibrium constant and the maximum useful work (the loss of free energy).

Let us consider the reaction $N_2 + 3H_2 \rightleftharpoons 2NH_3$. The reaction represents a reversible interaction between the components of a gaseous mixture of nitrogen, hydrogen, and ammonia. Let us place this mixture in a hermetically sealed box in which there are three apertures (Fig. 14). Aperture A can pass only nitrogen, aperture B only hydrogen, and aperture C only ammonia. Into each aperture is fastened a cylinder with a piston. The piston moves without friction, and the cylinder may be tightly closed where it is attached at the aperture. An equilibrium mixture of gases in the box has the partial pressures P_{N_2}, P_{H_2}, and P_{NH_3}. Apart from the box, we set up three balloons, each containing one of the gases (N_2, H_2, and NH_3) at pressures of P_1, P_2, and P_3 respectively. The balloons are sufficiently large so that the removal of a small quantity of gas has practically no effect on the pressure. Let us disconnect the cylinder and piston from aperture A and remove from the balloon with nitrogen one mole of gas. Let us close the cylinder and, by means of the piston, convert the one mole of nitrogen from the nonequilibrium reaction of pressure P_1 to the equilibrium pressure P_{N_2}. After this, introduce the gas into the reaction chamber through the aperture. The work of changing the gas from pressure P_1 to the equilibrium pressure P_{N_2} is expressed by the equation

$$A_1 = RT \ln \frac{P_1}{P_{N_2}}$$

A similar procedure is followed with three moles of hydrogen, as a result of which the work performed will be

$$A_2 = 3RT \ln \frac{P_2}{P_{H_2}}$$

After introducing into the chamber nitrogen and hydrogen, according to the law of mass action, the reaction

$$N_2 + 3H_2 \rightleftharpoons 2NH_3$$

will move toward the formation of ammonia. Let us remove through aperature C two moles of ammonia and convert it from the equilibrium reaction of pressure P_{NH_3} to the pressure P_3. In doing this the work performed will be

$$A_3 = 2RT \ln \frac{P_{NH_3}}{P_3}$$

As a result of the above effects we have obtained two moles of ammonia, having consumed one mole of nitrogen and three moles of hydrogen. The equilibrium relations of the gases in the chamber have remained undisturbed. The total work in obtaining two moles of ammonia proves to be

$$A_1 + A_2 + A_3 = RT \ln \frac{P_1}{P_{N_2}} + 3RT \ln \frac{P_2}{P_{H_2}} + 2RT \ln \frac{P_{NH_3}}{P_3}$$

$$A_{tot} = RT \ln \frac{P_1 \cdot P_2^3}{P_3^2} - RT \ln \frac{P_{N_2} \cdot P_{H_2}^3}{P_{NH_3}^2}$$

In these equations P_{N_2}, P_{H_2}, and P_{NH_3} represent the pressures when the reacting mixtures of $N_2 + 3H_2 \rightleftharpoons 2NH_3$ are in a state of equilibrium, since the work was performed each time for transformation from nonequilibrium to equilibrium pressure in the reacting mixtures and in the reverse direction in the reaction products. We may therefore write

$$\frac{P_{NH_3}^2}{P_{N_2} \cdot P_{H_2}^3} = K_p$$

where K_p is the equilibrium constant.

$$A_{tot} = RT \ln K_p - RT \ln \frac{P_3^2}{P_1 \cdot P_2^3} = -\Delta G$$

As we have already stated, the total work performed by a system under isothermal conditions defines the loss of free energy. In the present case we have shown the loss of free energy in going from an arbitrary constant (P_1, P_2, and P_3) to the equilibrium state. If $P_1 = P_2 = P_3 = 1$ atm, then $A_{tot} = -\Delta G = RT \ln K_p$. The value of ΔG applying to these conditions is written as ΔG^0, and the state when the partial pressure of each gas is equal to 1 atm at $t = 25°C$ is called the standard state. For a better understanding of the process let us once again trace the derivations. The change of N_2 and H_2 from arbitrary pressures to equilibrium pressures is accompanied by the work

$$A_1 = RT \ln \frac{P_1}{P_{N_2}}, \quad A_2 = 3RT \ln \frac{P_2}{P_{H_2}}$$

The change of NH_3 from equilibrium pressure of the reaction to nonequilibrium pressure is characterized by the work of the system

$$A_3 = 2RT \ln \frac{P_{NH_3}}{P_3}$$

when $P_1 = P_2 = P_3 = 1$ atm, then $A_1 = -\Delta Z_{N_2}^0$; $A_2 = -3\Delta Z_{H_2}^0$; $A_3 = 2\Delta Z_{NH_3}^0$, (the last ΔG^0 has the opposite sign since the process is opposite to the first two).

$$A_1 + A_2 + A_3 = \Delta G_{N_2}^0 - 3\Delta G_{H_2}^0 + 2\Delta G_{NH_3}^0 = RT \ln \frac{P_{NH_3}^2}{P_{N_2} \cdot P_{H_2}^3}$$

$$-\Delta G_{reaction}^0 = 2\Delta G_{NH_3}^0 - \left(\Delta G_{N_2}^0 + 3\Delta G_{H_2}^0\right) = RT \ln K_p$$

$$K_p = \exp\left(-\frac{\Delta G_{reaction}^0}{RT}\right)$$

We may thus draw the following conclusion: to determine the equilibrium constant it is necessary to subtract from the standard change in free energy of the final products the standard changes in free energy of the reacting substances. Since tabular values of the standard changes in free energy are given for a single mole, ΔG^0 must be taken according to the stoichiometric coefficients of the reacting substances and the reaction products. As is well known, for infinitely dilute solutions the assumption of partial pressures is entirely applicable: the concentration of a certain ion in water may be correlated with the partial pressure. We may therefore write

$$A_{tot} = -\Delta G_{reaction}^0 = RT \ln K$$

Thus, in comparing $-\Delta G^0$ of different substances we determine how far they are from the equilibrium state at concentrations of each mole per liter or partial pressure of 1 atm. If we go from an infinitely dilute solution to an actual solution, then, relative to each concentration, it is necessary to take a correction into account, the activity coefficient, or to use the activities of the ions.

The equation we have obtained makes it possible to compute the solubility product or the equilibrium constant of any reaction if its ΔG^0 is known. Actually, since $\Delta G^0 = -RT \ln K$, then, for the condition $T = 273.16 + 25°C$ and after converting natural logarithms to those on the base ten, we may write

$$\Delta G^0 = -1.364 \lg K$$

$$K = \text{antilog}\left(-\frac{\Delta G_{reaction}^{\bullet}}{1.364}\right)$$

or, if the final products are in the solid phase

$$SP = \text{antilog}\left(-\frac{\Delta G_{reaction}^{\bullet}}{1.364}\right)$$

On the basis of the above equation we determine the solubility product of iron hydroxide:

$$Fe(OH)_3 = Fe^{\cdots} + 3(OH)'$$

The change (diminution) in free energy will be determined by the difference

$$\left(\Delta G_{Fe^{\cdots}}^0 + 3\Delta G_{(OH)'}^0\right) - \Delta G_{Fe(OH)_3}^0$$

Having selected the appropriate values of free-energy increment from the table (see Appendix), we find

$$(-3.00 - 112.78) + 169.45 = 53.67$$

$$SP = \text{antilog}\left(-\frac{53.67}{1.364}\right) = 4.5 \cdot 10^{-40}$$

This method of computation will be used extensively in the present book.

The Concept of the Oxidation – Reduction Potential

Any work performed may be expressed as the product of two factors. For example, the mechanical work is determined by the product of the force and the path distance along which the applied force acted:

$$A = F \cdot S$$

The force is an intensity value characterizing the directivity of the process, and the distance over which the force is applied is a factor bearing the name extensiveness, or capacity. The work of an expanding gas in a cylinder at constant pressure is defined as the product of the pressure and the volume difference:

$$A = P(V_2 - V_1)$$

Here, the intensity factor is pressure, acting on the piston of the cylinder, and the capacity factor is the volume change. The work of the earth's gravitational force during free fall is expressed by mgh, where gh characterizes the intensity of the process and is the intensity factor, and m is the mass on which the intensity factor acts. The latter is the extensiveness (capacity) factor. The work of an electrical current is determined by the product of the electromotive force and the amount of electricity transferred:

$$A = E \cdot F$$

where the electromotive force is the intensity value, characterizing the strength or intensity of the process, and the transferred amount of electricity is the capacity factor (or extensiveness).

In examining these examples it is not difficult to observe the general characteristic found in the concept of work: any form of work may be represented by two factors, intensiveness and extensiveness. By virtue of its specific character, the factor of intensiveness or intensity appears only where there exists a difference in energy levels as a minimum of two systems. The process will take place in a direction to eliminate this difference. The free fall of a body takes place only where there is a difference of potential energy. Current will pass through a conductor only when there is a potential difference between the ends of the conductor. Gas in a cylinder will move the piston only when there is a difference between external and internal pressures or a temperature difference between the heat source and the active body.

The cited examples should impress the mind with the difference between intensive and extensive values: intensive values do not obey the law of additivity, whereas extensive values do. For example, the volume (an extensive value) of any system may be subdivided into individual parts. Each part proves to be proportional to the entire volume. The sum of all the parts is therefore equal to the whole. It is impossible to speak of temperature as an intensive value. If any body is subdivided into parts, the temperature of each part remains the same as that characteristic of the whole body. The temperature of a body will not be equal to the sum of the temperatures of the individual parts.

This important distinction between the factors specifies the conditions for measuring them. The extensive characteristic of an entire body may be obtained by selecting any part of the body as a unit and by measuring this selected unit. Thus, to measure distance we use the unit of length the centimeter (an extensive value), and for measuring volume we select a $1 \cdot cm^3$ volume. The volume of any system is defined by a number indicating how many units of volume contained in the entire system, and so forth.

Since an intensive value does not obey the law of additivity, the characteristic of such a value may be expressed only in comparison with a similar intensive factor of another system. We naturally lose the chance of determining any absolute value of the intensive factor and are satisfied merely with a relative value: the degree to which one system has a greater (or smaller) energy potential in comparison with another. For example, when we speak of pressure, we have in mind a comparative value relative to the pressure of the atmosphere at sea level and a latitude of 45°. In this case the standard of comparison is called one normal atmosphere. When we speak of force, we have in mind the change in

amount of movement with time relative to a standard mass. The relativity of an intensive value may be very well grasped from the example with temperature.

If we take two bodies at different temperatures and place them in contact, the energy levels between the two begin to adjust; the warmer body will yield heat to the cooler. In other words, energy of the body with higher temperature is transferred to the body having lower temperature. In order to determine the direction in which the heat exchange will take place it is not necessary to know the absolute temperatures of these bodies. It is only necessary that a temperature difference exists between the bodies for a directed exchange of energy to take place.

A good possibility for comparing thermal energy levels appears when we use for our determinative standard an easily reproducible amount of heating. The process of selecting a standard degree of heating that may be rather accurately reproduced has a rather long history. Different states in different systems have been proposed for the standard or fiducial point of heating. At one time it was even suggested that the melting point of butter be used as the standard. Fahrenheit first proposed the temperature of a mixture of snow and common salt and also the temperature of the human body as standards.

Even after it was shown that the most convenient standard, from the viewpoint of reproducibility, is the degree of warming to melt ice, there were wide discussions concerning the value to assign to this degree of warming. Mendeleev proposed that this be given the value of 1000°, Celsius 100°, and Stromer 0°. At the present time the standard for degree of warming in many countries has been taken as the melting point of ice and has been designated 0°. The second point on the thermometer is the temperature at which water boils at an external pressure of one normal atmosphere.

This example must show that the choice of a standard energy level for comparing intensive values is rather arbitrary and the number assigned to it may be arbitrary. However, having chosen some degree of heat as a standard and having assigned it a definite symbol, and having marked off, above and below this point, divisions corresponding to other degrees of heat, we may always obtain comparable results for the energy characteristics of all systems. It is important, once again, to emphasize that 0°C is a temperature conditionally adopted, characterizing a definite degree of heat. Every body with a temperature below 0°C is characterized by a lower energy level, the value being marked by a minus sign.

In this case it becomes unnecessary to compare bodies among themselves to discover the direction of energy exchange. It is sufficient to know what temperature they have relative to a standard point. Thus, if two bodies have different temperatures, one $-5°$, and the other $-15°$, the first will be "warm," the second "cold," and the exchange of energy will take place from the first to the second.

If the bodies have identical masses and heat capacities, the exchange of energy will cease when the temperature of both bodies is $-10°$.

We have described in detail an example involving temperature, since the oxidation—reduction potential, as will become clear below, is also an intensive value, and the entire discussion concerning temperature may be applied word for word to Eh. The analogy with temperature is especially important in discussing the problem of searching for the "neutral point" of the oxidation—reduction potential. Having gained an idea concerning intensive and extensive values, we may proceed to an investigation of the problem of representing the oxidation—reduction potential of a system.

Let us assume that we have an aqueous solution of ferric- and ferrous-oxide salts. No restrictions will be placed on this system: the activities of both ferric and ferrous ions may be arbitrary. Since the ions differ in valence, the following equilibrium must exist between them: $Fe^{\cdot\cdot} = Fe^{\cdot\cdot\cdot} + e$. As for all neutral reactions, we must determine the equilibrium constant

$$K = \frac{a_{Fe^{\cdot\cdot\cdot}} \cdot a_e}{a_{Fe^{\cdot\cdot}}}$$

But, whereas the equilibrium constant for neutral reactions defines only the ionic ratio, in the given case the ionic ratio depends also on the electrostatic state of the system. It is therefore more convenient to write

$$\frac{a_{Fe^{\cdot\cdot\cdot}}}{a_{Fe^{\cdot\cdot}}} = K/a_e$$

where a_e is the activity of the electrons, characterizing the value of the charge, the function of which is the given equilibrium state $a_{Fe^{\cdot\cdot\cdot}}/a_{Fe^{\cdot\cdot}}$. After making logarithmic transformations,

$$\ln K = \ln \frac{a_{Fe^{\cdot\cdot\cdot}}}{a_{Fe^{\cdot\cdot}}} + \ln a_e$$

But, since $-\Delta G^0_{reaction} = RT \ln K$, for the given electrochemical reaction

$$-\Delta G^0_{Fe^{\cdot\cdot\cdot}/Fe^{\cdot\cdot}} = RT \ln \frac{a_{Fe^{\cdot\cdot\cdot}}}{a_{Fe^{\cdot\cdot}}} + RT \ln a_e$$

As we know, $-\Delta G^0$ represents the standard change in free energy as applied to one mole or one gram-ion. This standard change in free energy characterizes the work performed by a system in going from the state where $a_{Fe^{\cdot\cdot\cdot}} = a_{Fe^{\cdot\cdot}} = 1$ gram-ion to the state of equilibrium ion content $a_{Fe^{\cdot\cdot\cdot}}/a_{Fe^{\cdot\cdot}}$ determined by the given electrostatic characteristics. The electrostatic characteristics may be defined by the potential of the given system. For this, we recall that the work of an electrical current is determined by the product of the potential difference and the amount of transferred electricity: $A = EF$. The potential difference is the difference of intensive values. Since $A = -\Delta G^0$,

$$E \cdot F = RT \ln \frac{a_{Fe^{\cdot\cdot\cdot}}}{a_{Fe^{\cdot\cdot}}} + RT \ln a_e$$

or

$$E_1 - E_2 = \frac{RT}{F} \ln a_e + \frac{RT}{F} \ln \frac{a_{Fe^{\cdot\cdot\cdot}}*}{a_{Fe^{\cdot\cdot}}}$$

where E_1 is the potential of the standard state of the system (i.e., of the state when $a_{Fe^{\cdot\cdot\cdot}} = a_{Fe^{\cdot\cdot}} = 1$ gram-ion), E_2 is the potential of the equilibrium system corresponding to the given electrostatic state, and F is the charge per valence unit in 1 gram-ion.

Thus, at any $a_{Fe^{\cdot\cdot\cdot}}/a_{Fe^{\cdot\cdot}}$ ratio,

$$E_2 = E_1 - \left(\frac{RT}{F} \ln a_e + \frac{RT}{F} \ln a_{Fe^{\cdot\cdot\cdot}}/a_{Fe^{\cdot\cdot}} \right)$$

E_1 is always constant, since it refers to the standard state, but E_2 may be changed arbitrarily in connection with a change in the charge and, correspondingly, to a change in $a_{Fe^{\cdot\cdot\cdot}}/a_{Fe^{\cdot\cdot}}$. We cannot measure the absolute values of either E_2 or E_1, since they are intensive values. And, without knowing the value of E_2, we cannot investigate the problem of how the ferric—ferrous system will behave when connected to some other system such as $Mn^{\cdot\cdot} \rightarrow Mn^{\cdot\cdot\cdot}$ having a potential of E_2'. We cannot state, without having connected these systems directly to each other, whether $Mn^{\cdot\cdot}$ or $Fe^{\cdot\cdot}$ will be oxidized, since we do not know what sign $E_2 - E_2'$ may have. In order to make such an evaluation, without conducting direct measurements, it would be necessary to arrange a standard system, in relation to which we might determine the potentials of all other systems. Then, in comparing any systems having quantitative evaluation relative to the standard, we shall be able immediately to speak of the direction of the process in the same way we were able to determine the direction of heat exchange without going directly to the contact between the bodies, merely by knowing the temperatures. As a system to use for comparison, let us use a solution with a pH of zero, through which shall pass gaseous hydrogen at a pressure of 1 atm. If the potential of this system is assumed to be zero, then, as we shall see below, the computation of all other systems takes on a very simple form.

Thus, in the comparative system an electrochemical reaction takes place:

$$\tfrac{1}{2}H_2 \rightleftarrows H^{\cdot} + e, \quad K/a_e = \frac{a_{H^{\cdot}}}{P_{H_2}^{1/2}}$$

As for the ferric—ferrous system, we find

$$E^0 = E_1^0 - \left(\frac{RT}{F}\ln a_e + \frac{RT}{F}\ln \frac{a_{H^{\cdot}}}{P_{H_2}^{1/2}} \right)$$

where E_1^0 is the potential of the system at $a_{H^{\cdot}} = 1$ gram-ion/liter, $P_{H_2} = 1$ atm, and E^0 is the potential of the system under equilibrium conditions corresponding to some electron activity. Since we have been given the conditions that $E_1^0 = 0$; $a_{H^{\cdot}} = 1$ gram-ion/liter, and $P_{H_2} = 1$ atm,

$$\frac{RT}{F}\ln \frac{a_{H^{\cdot}}}{P_{H_2}^{1/2}} = 0 \quad \text{and} \quad E^0 = -\frac{RT}{F}\ln a_e$$

In this case the potential difference between the solution with arbitrary ferric- and ferrous-ion activity and the standard system of hydrogen is determined from the equation

$$E_2 - E^0 = E_1 + \frac{RT}{F}\ln \frac{a_{Fe^{\cdots}}}{a_{Fe^{\cdot\cdot}}}$$

whence

$$E_1 - E_2 = -\left(E^0 + \frac{RT}{F}\ln \frac{a_{Fe^{\cdots}}}{a_{Fe^{\cdot\cdot}}} \right)$$

It is easy to determine E^0 from the last equation. For this we rewrite the equation in the following form:

$$\left(E^0 \cdot F + RT \ln \frac{a_{Fe^{\cdots}}}{a_{Fe^{\cdot\cdot}}} \right) = -(E_1 - E_2) \cdot F = -A = \Delta G^0_{Fe^{\cdot\cdot}/Fe^{\cdots}}$$

With the condition $a_{Fe^{\cdots}} = a_{Fe^{\cdot\cdot}}$,

$$E^0 = \frac{\Delta G^0_{Fe^{\cdot\cdot}/Fe^{\cdots}}}{F}$$

Thus, E^0 represents the potential difference between the standard system of ferric and ferrous ions (i.e., the system where $a_{Fe^{\cdots}} = a_{Fe^{\cdot\cdot}} = 1$ gram-ion) and the standard hydrogen system. The value of E^0 is called the standard potential of the oxidation—reduction system relative to the standard hydrogen electrode. After representing $(E_1 - E_2)$ by

Eh for the ferric—ferrous system, we write

$$Eh = E^0 + \frac{RT}{F}\ln \frac{a_{Fe^{\cdots}}}{a_{Fe^{\cdot\cdot}}}$$

This equation was derived on the basis of the ferric—ferrous system. But, in general, for any system, $B^{\nu} = B^{\eta} + n$, where $n = (\eta - \nu)$ electrons,

$$Eh = E^0 + \frac{RT}{Fn}\ln \frac{B^{\eta}}{B^{\nu}}$$

where E^0 is the standard potential of the given system relative to the normal hydrogen electrode.

In this case

$$E^0 = \frac{\Delta G^0_{B^{\nu}/B^{\eta}}}{Fn}$$

F = 96,484 absolute coulombs. Since ΔG^0 in the table is generally given in kilocalories, F is better expressed in kilocalories per volt, referred to one mole or gram-ion. This amounts to 96,484 · 4184 J · 10^{-3} = 23.06. Thus

$$E^0 = \frac{\Delta G^0_{reaction}}{23.06 \cdot n}$$

where n is the number of valence electrons participating in the reaction.

For t = 25°C, RT/Fn · ln x = 0.059/n lg x. Therefore, for the ferric—ferrous system under consideration, we may write

$$Eh = E^0 + 0.059 \lg \frac{a_{Fe^{\cdots}}}{a_{Fe^{\cdot\cdot}}}, \quad E^0 = \frac{\Delta G^0_{Fe^{\cdots}} - \Delta G^0_{Fe^{\cdot\cdot}}}{23.06}$$

Using the table (see Appendix), we find

$$Eh = 0.75 + 0.059 \lg \frac{a_{Fe^{\cdots}}}{a_{Fe^{\cdot\cdot}}}$$

where $a_{Fe^{\cdots}}$ and $a_{Fe^{\cdot\cdot}}$ represent the forms of ions in solution.

We have derived this equation on the basis of possible arbitrary activity of the ferrous and ferric ions in the solution. In practice this case is possible under limited conditions, by no means for all previously devised concentrations. The values of $a_{Fe^{\cdots}}$ and $a_{Fe^{\cdot\cdot}}$ may range between wide limits only under acid conditions. With increase in pH hydrolysis becomes stronger and at some excess of $a_{Fe^{\cdots}}$ ferric hydroxide begins to participate.

Let us consider how we shall determine the oxidation—reduction potential of the system in this case. We shall rewrite the equation of the oxidation—reduction state of the ferric—ferrous system in the following form

$$Eh = E^0 + 0.059 \lg a_{Fe^{\cdots}} - 0.059 \lg a_{Fe^{\cdot\cdot}}$$

When hydrolysis of the ferric ion begins, the activity of this ion is under control in the solution, being strictly determined by the pH value. We have already spoken of this at the beginning of the chapter, having shown that the control in this case is determined by the reaction

$$Fe^{\cdots} + 3(OH)' = Fe(OH)_3$$

whence (see p. 19)

$$a_{Fe^{\cdots}} = 4.5 \cdot 10^2 \cdot a_H^3.$$

or, in the logarithmic form,

$$\lg a_{Fe^{\cdots}} = 2.65 - 3pH$$

Let us substitute the value obtained for $\lg a_{Fe}\ldots$ in the equation of the oxidation—reduction potential. Then

$$Eh = E^0 + 0.157 - 0.177pH - 0.059 \lg a_{Fe^{\cdots}}.$$

After substituting the numerical value of E^0, we find

$$Eh = 0.907 - 0.177pH - 0.059 \lg a_{Fe^{\cdots}}. \qquad (6)$$

Let us determine what reaction it is that Eq. (6) corresponds to. To determine $\lg a_{Fe}\ldots$ we shall start with Eq. (1), which corresponds to the reaction

$$Fe(OH)_3 = Fe^{\cdots} + 3(OH)'$$

Taking into account the dissociation of water, $a_{(OH)'}$ was replaced by a_H. in keeping with the reaction

$$H^{\cdot} + (OH)' = H_2O$$

Thus, $\lg a_{Fe}\ldots = 2.65 - 3pH$ corresponds to the reaction

$$+ \quad \begin{array}{l} Fe(OH)_3 = Fe^{\cdots} + 3(OH)' \\ 3H^{\cdot} + 3(OH)' = 3H_2O \\ \hline Fe(OH)_3 + 3H^{\cdot} = Fe^{\cdots} + 3H_2O. \end{array}$$

The value of $\lg a_{Fe}\ldots$ obtained was introduced into the oxidation—reduction equation of the ferric-ferrous system, describing the reaction

$$Fe^{\cdot\cdot} = Fe^{\cdots} + e$$

In other words, the following summation was produced:

$$+ \quad \begin{array}{l} Fe^{\cdots} + 3H_2O = Fe(OH)_3 + 3H^{\cdot} \\ Fe^{\cdot\cdot} = Fe^{\cdots} + e \\ \hline Fe^{\cdot\cdot} + 3H_2O = Fe(OH)_3 + 3H^{\cdot} + e. \end{array} \qquad (7)$$

Knowing the general rule for determining the oxidation—reduction potential, let us check to see if reaction (7) corresponds to Eq. (6):

$$\Delta G^0_{Fe(OH)_3} + 3\Delta G^0_{H^{\cdot}} - (\Delta G^0_{Fe^{\cdot\cdot}} + 3\Delta G^0_{H_2O}) = \Delta G^0_{reaction}$$

Let us substitute the appropriate values of ΔG^0, taken from the Appendix:

$$-169.45 + 20.30 + 170.07 = 20.92$$
$$E^0 = \frac{20.92}{23.06} = 0.907$$

Equation (7) thus corresponds to the oxidation—reduction Eq. (6). One's attention is drawn to the fact that in Eq. (6), characterizing the Eh of the solution, only pH and $a_{Fe}\ldots$ appear as variables, i.e., elements of the solution. Eh is not defined by the amount of iron hydroxide present in the solid phase, although this compound participates in the reaction. In the given case we find an analogy with the solubility product, when the activities of the solid phases do not enter the expression of the equilibrium constant. Therefore, essentially for any reaction of the type

$$A^m + B + N^{\nu} = A^n + D + C^{\lambda} + (n - m) e$$

we may write

$$Eh = E^0 + \frac{0.059}{n - m} \lg \frac{a_{A^n} \cdot a_{C^{\lambda}}}{a_{A^m} \cdot a_{N^{\nu}}}$$

where A^n and A^m are ions in the oxidized and reduced forms, respectively; B and D are solid phases; N^{ν} and C^{λ} are ions of the solution not participating in the oxidation—reduction reactions.

In the general case the actual oxidation—reduction potential of any reversible reaction (in the electrochemical sense) at 25°C may be computed from the equation

$$Eh = E^0 + \frac{0.059}{n} \lg \frac{oxidized\ form}{reduced\ form}$$

where by oxidized form we mean all ions in the solution participating on the right side of the reaction, and by reduced forms we mean all ions in the solution on the left of the equal sign. This equation is called the Nernst equation, and it represents one of the basic equations that will be used in the present book. We therefore think it necessary to emphasize the essential details that must be utilized in deriving this equation.

1. The oxidation—reduction potential defines not the work but only the trend and intensity of the process, since it is an intensity factor. Consequently, we cannot determine the value of the oxidation—reduction potential from the amount of reduced material.

2. Being an intensive factor, the oxidation—reduction potential characterizes only the potential relative to the standard hydrogen electrode. If Eh = 0, this does not mean any "stabilization" of

the oxidation and reduction processes or any "neutral environment," but it merely indicates that the given system has the same potential as the standard hydrogen electrode.

3. The oxidation–reduction potential reflects the equilibrium of ions in solution. The ions are in equilibrium with the solid phase (if it exists) but generally not in equivalent quantities. The oxidation–reduction potential in no case depends on the amount of the solid phase.

These conclusions, following from the very essence of the concept of an oxidation–reduction potential of a medium, may make it possible to evaluate the methodological accuracy of a construction based on comparison of the oxidation–reduction potential of an environment with the ratio of the forms of iron or sulfur in the solid phase in the sediment.

Analysis for Some Errors

Let us look at a somewhat simplified but still rather clear example. Iron hydroxide is carried into a basin. Because of definite reducing conditions in the mud, the following reaction should take place:

$$Fe(OH)_3 + 3H^{\cdot} + e = Fe^{\cdot\cdot} + 3H_2O$$

With the oxidation–reduction potential existing in the mud, equilibrium of this system will be determined by the equation

$$Eh = E^0 - 0.177pH - 0.059 \lg a_{Fe^{\cdot\cdot}}.$$

Depending on Eh and pH, $a_{Fe^{\cdot\cdot\cdot}}$ should take on different values, possibly reaching such a degree of saturation that iron will precipitate out as $Fe(OH)_2$. For this to happen, a low oxidation–reduction potential would be necessary, lower than any value thus far observed in nature. The precipitation of $Fe(OH)_2$ for the Eh range existing in nature must therefore be excluded. But, in analyzing a 5%-HCl extract of mud, only $Fe^{\cdot\cdot\cdot}$ will be detected, regardless of the oxidation–reduction potential of the medium, since $Fe(OH)_3$ dissolves rather easily in hydrochloric acid. This example helps us to understand the lack of firm grounds for comparing $Fe^{\cdot\cdot}/Fe^{\cdot\cdot\cdot}$ with Eh or Eh with the form of sulfur in the sediment.

It may be said, of course, that the example is trivial, that in nature everything is more complex and a whole series of other minerals forms in addition to iron hydroxide. All this is true, but objections of this kind nevertheless do not furnish proof of suitability of comparing Eh with elements of

different valence in the sediment. Still, we do not insist that the objection we have noted concerning the behavior of iron hydroxide at different values of Eh exhausts the list.

Errors of investigations that compare the forms or iron or sulfur in sediments with the Eh are more veiled, although essentially they reveal a series of similar equations and, in the process of further elucidation of the problem, they stand out in all fullness. Here we shall pause to consider some errors that directly contradict the above conclusions concerning oxidation–reduction potential, though this circumstance unfortunately does not slow down the use of the erroneous results. The most outstanding violation in use of the Nernst equation has been noted in the works of Savich [1950, 1956].

Thus, Savich wrote [1956] that "the thermodynamic equation correlating the value of the Oxr† potential with concentrations of ferric and ferrous oxides has the form

$$Eh = E^\circ + \frac{RT}{F} \ln \frac{[Fe^{\cdot\cdot\cdot}]}{[Fe^{\cdot\cdot}]}$$

Substituting numerical values for R, T, F, and E^0 and replacing natural logarithms by logarithms to the base ten, we obtain

$$Eh = 0.677 + 0.058 \lg \frac{[Fe^{\cdot\cdot\cdot}]}{[Fe^{\cdot\cdot}]}$$

Whence, knowing the content of ferric and ferrous iron in the sediments, we may compute the value of Eh" (emphasis mine – M. S.). Having said this, the author proceeded by noting that he substituted in the Nernst equation the value of iron extracted by a 5%-HCl extract! This very clear mistake has passed unnoticed, and the data obtained by Savich (naturally incorrect) have been widely used in the geologic literature as a proof for the correlation between form of iron in the sediment and the oxidation–reduction potential of the medium [Strakhov, 1954, 1959]. It is claimed that Savich has explained the pattern in the relations between the ratio of forms of iron in the sediment and the oxidation–reduction potential of the solution [Romankevich and Petrov, 1961]. Following Savich, Vainbaum and Il'inskaya [1960] considered it possible to analyze the ratio of oxidized and reduced forms of iron in a rock and substitute the analytical results in the Nernst equation. We have shown the results obtained by these investigators (Table 3).

_____
†In his work, Savich designates this "oxidation–reduction."

TABLE 3. Comparison of Measured and Computed Values (after Vainbaum and Il'inskaya [1960])

Sample locality	Lithology	Eh measured	Eh computed	Eh accounting for hydrolysis*	pH	p Fe··	p Fe···
Zol'nyi Ovrag, bh 65, 540.8 m	Clay	+0.575	+0.690	+0.630	3.2	1.31	1.01
Obmarovka, bh 1, 1159.0 m	Siltstone	+0.349	+0.770	+0.331	5.25	2.27	0.59
" " 1169.8 m	Clay	+0.577	+0.629	+0.777	2.2	0.83	1.59
" " 1644.0 m	Siltstone	+0.511	+0.713	+0.523	4.3	2.75	2.05
Mukhanovo, bh 18, 2846.7 m	Clay	+0.496	+0.746	+0.520	4.1	2.14	0.87
" " 2875.0 m	Limestone	+0.335	+0.711	-0.173	7.8	1.24	0.58

*Without consideration of hydrolysis, the equation used was: $Eh = E^0 + 0.059 \lg[Fe^{···}/Fe^{··}]$. Taking hydrolysis into account, the equation was $Eh = E^0 - 0.177\ pH - 0.059\ \lg Fe^{··}$.

From Table 3 it is clearly seen that the computed and measured values of Eh do not agree. The authors of the work also noted this, but they referred the discrepancies to the antiquity of the sediments, not suspecting that they were comparing values that should not be compared. Similar errors were found in the works of Yurganov [1956a, 1956b] and Poddubnyi [1959]. The intuitive assumption on the basis of which the oxidation–reduction potential of a medium is correlated with the ratio of the forms of iron in the sediment has led to investigations along spurious paths [Ontoev, 1956]. Such work naturally cannot serve as confirmation of the investigated correlation. It is merely based on the a priori assumption that this correlation must exist, but, as we have shown, the assumption has not been proved valid.

Since all values of Eh are referred to the hydrogen electrode, positive values of Eh mean that a given system has a stronger oxidizing potential than the standard hydrogen system, and a negative value indicates that a given system is a reducing one as compared with the standard hydrogen system. Consequently, the standard hydrogen system will reduce a system with Eh > 0 and oxidize a system with Eh < 0. If one system has Eh > 0 and another Eh ≫ 0, when these two systems are combined, the second will oxidize the first. Since, depending on the activities of the components in solution, Eh of a particular system may range between rather wide limits, standard potentials are generally compared (E^0), i.e., conditions such that the oxidized and reduced forms are present in equal quantities.

An example of such a comparison is shown in Fig. 15, in which the normal potentials of metal-cations are plotted in relation to the standard hydrogen electrode. If the indicated pairs are found in solution, in keeping with the requirements of the standard potential, each lower-lying pair is reduced by the preceding. Thus, if E^0 or Eh is zero, this means that the oxidation–reduction processes become stabilized, that only reduction takes place at values below zero, and only oxidation at values above zero. In this case the term "neutral oxidation–reduction potential," introduced into the geological literature, is improper. The term has been used frequently in the works of Teodorovich. Just how very fine this term sounds becomes clear from the equivalent concept of "neutral temperature," referred to 0°C.

There is no doubt that it is through such inaccuracies, which would seem hardly to deserve attention, that fundamental physicochemical con-

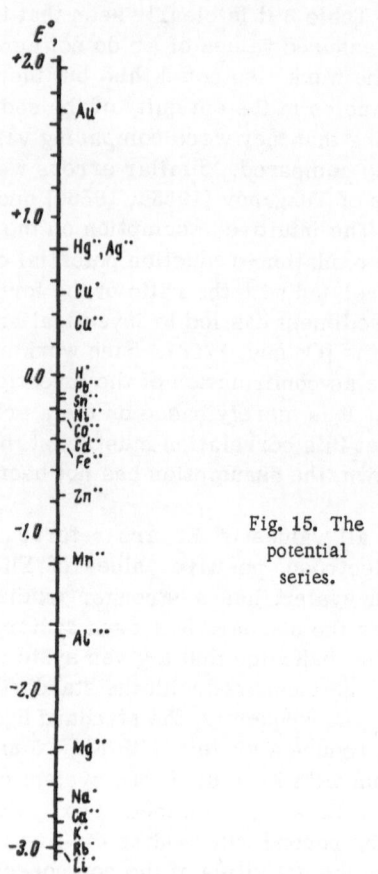

Fig. 15. The potential series.

cepts gradually begin to undergo distortions. It seems to us that such inaccuracies may explain the error that an increase in oxidation—reduction potential downward in modern sediments, with the value approaching zero, must indicate that the oxidation—reduction processes approach the equilibrium state, i.e., neutral conditions [Gordeev, 1962].

The question arises: can we speak, in general, of a transitional zone between oxidizing and reducing conditions? From the above discussion it is clear that each pair of ions has its "neutral point," reflecting the potential E^0, since only at this potential in the solution is the activity of the oxidized form equal to the activity of the reduced form. But for another element E^0 will have a different value. Therefore, generally speaking, there is no objectively based beginning or end of the scale, nor any objective mean or neutral point [Mikhaélis, 1932]. This, of course, creates a certain diffuseness in the concept of oxidizing and reducing conditions, as may be seen from the following examples.

Shcherbina [1949] proposed that we use, for the transition from oxidizing to reducing conditions, the conditions at which 50% of the iron is in the oxi-

dized form, 50% in the reduced form. Bushinskii [1954] considered the value Eh = 200 mV to represent the lower boundary of the oxidizing environment. Gulyaeva [1955] proposed the Eh values of −508 and −610 mV for the transitional boundary between oxidizing and reducing conditions. Shcherbakov [1956], on the basis of a study of groundwater, concluded that oxidizing conditions must be characterized by values of Eh ≫ 250 mV, oxidation—reduction conditions by Eh = 250−0, reducing conditions by Eh = 0 − (−150) mV, and so forth. Sokolova [1961] referred to the zone of Eh = 180 mV as transitional to weakly reducing conditions.

All the complexity in choosing transitional conditions lies in the fact that the oxidation—reduction potential characterizes an intensive factor of the process and, consequently, reflects a relative concept. The best solution to the problem, therefore, would be to select those criteria that would reflect some particular oxidation—reduction environment, always constant, and that would be rather widespread in nature. All the criteria listed in the first chapter would serve this purpose, but their application has unfortunately proved not altogether successful.

Iron minerals are most widely distributed in sedimentary rocks in comparison with other authigenic minerals. They have attracted investigators as principal indicators of physicochemical conditions. We shall try to shed light on the role of these minerals for purposes of establishing oxidation—reduction conditions. A warning must be issued, however, that even should such a characteristic appear for any mineral, this does not mean that, in conditions more highly oxidizing than those under which the mineral formed, only oxidizing processes will take place, or that in a more reducing environment, only reducing processes will take place. We shall demonstrate this later in regard to the solid phase, but here we shall consider it in relation to the ionic state of a solution.

For the system $Fe^{\cdot\cdot} = Fe^{\cdot\cdot\cdot} + e$,

$$Eh_1 = 0.75 + 0.059 \lg \frac{a_{Fe^{\cdot\cdot\cdot}}}{a_{Fe^{\cdot\cdot}}}$$

for the system $Mn^{\cdot\cdot} = Mn^{\cdot\cdot\cdot} + e$,

$$Eh_2 = 1.51 + 0.059 \lg \frac{a_{Mn^{\cdot\cdot\cdot}}}{a_{Mn^{\cdot\cdot}}}$$

and, finally, for the system $\frac{1}{2} H_2 = H^{\cdot} + e$,

$$Eh_3 = 0.059 \lg \frac{a_{H^{\cdot}}}{P_{H_2}^{1/2}}$$

If the three systems are found in a single solution, this solution must have some single oxidation—reduction potential. Therefore, $Eh_1 = Eh_2 = Eh_3$ and, consequently, the following ionic ratios obtain in the solution:

$$0.75 + 0.059 \lg \frac{a_{Fe^{\cdots}}}{a_{Fe^{\cdot\cdot}}} = 1.51 + 0.059 \lg \frac{a_{Mn^{\cdots}}}{a_{Mn^{\cdot\cdot}}} = 0.059 \lg \frac{a_{H^{\cdot}}}{P_{H_2}^{1/2}}$$

$$\frac{10^{12,7} \cdot a_{Fe^{\cdots}}}{a_{Fe^{\cdot\cdot}}} = \frac{10^{25,6} \cdot a_{Mn^{\cdots}}}{a_{Mn^{\cdot\cdot}}} = \frac{a_{H^{\cdot}}}{P_{H_2}^{1/2}}$$

If in this solution it is found that $a_{Fe^{\cdots}} = a_{Fe^{\cdots}}$, then for a single ion of $Mn^{\cdots}$ there will be approximately 10^{13} ions of $Mn^{\cdot\cdot}$. A neutral environment for iron proves to be a reducing one for manganese. A strong belief in a "neutral point" reconciles geologists with the following widespread but erroneous opinion.

"As physical chemistry teaches, the following relations are applicable to a medium in which dynamic physicochemical equilibrium has been established:

$$\frac{[Fe^{\cdots}]}{[Fe^{\cdot\cdot}]} = \frac{[Mn^{\cdots}]}{[Mn^{\cdot\cdot}]} = \frac{[S]}{[S'']} = \frac{[x^{\cdots}]}{[x^{\cdot\cdot}]} = \frac{[H^{\cdot}]}{[H]}$$

In other words, the ratio of amounts of higher and lower oxidized forms of a substance, when expressed in moles, will be the same for all existing pairs" [Savich, 1950, p. 143]. It is to the point that this same expression is found, word for word, in the 1936 work of Uspenskii [1936]. After almost 20 years, the error of microbiologists was transferred to the geologic literature, and it has appeared afresh even in works of later years [Sokolova, 1962; Serdobol'skii, 1965].

In conclusion, let us consider the basis of electrometrical methods of measuring the oxidation—reduction potential. The standard hydrogen electrode, the potential of which is taken as zero, is not very convenient in practical work because of its variability and the formation of diffusion potentials at places where electrode liquids are in contact. In practice, therefore, we most frequently use a calomel electrode for comparison. This is a reversible electrode, always having a constant potential relative to the standard hydrogen electrode and, therefore, always representing its own kind of null value, a distinctive standard of comparison.

The operating principle of the calomel electrode is the following. When we pour water on metallic mercury, a potential jump arises at the metal-water boundary. This follows from the fact that ions of mercury, torn from the metal, begin to move across into the layer of water above the mercury. In this process the metallic mercury is charged negatively, the solution above the mercury positively. The electrostatic attraction arising because of the difference in charges between the ions that have gone into solution and the ions in the oppositely charged mercury interrupts this process at a certain stage, and dynamic equilibrium is established in the system. If we replace the water above the mercury by a saturated solution of calomel (Hg_2Cl_2), the ions of mercury that appear because of dissociation of the calomel begin to pass into the metallic mercury, charging it positively. The solution itself is then charged negatively. This process cannot proceed for a protracted interval since the positively charged mercury will attract negatively charged chlorine ions to its surface, which will eventually isolate the metallic mercury to such an extent that the rate of mercury ions arriving at the metal and the rate of those departing will become equal.

The mercury thus acquires a certain potential because of the following reaction

$$Hg \rightleftharpoons Hg^{\cdot} + e$$

The value of this potential relative to the normal hydrogen electrode is determined by the Nernst equation:

$$Eh = E^0 + 0.059 \lg a_{Hg^{\cdot}}.$$

Here $a_{Hg^{\cdot}}$ is determined by the activity of the chlorine ions, since these values are interrelated by the solubility product of calomel (we used a saturated solution of calomel).

$$a_{Hg^{\cdot}} \cdot a_{Cl'} = K$$

whence

$$a_{Hg^{\cdot}} = \frac{K}{a_{Cl'}}$$

The oxidation—reduction potential is then

$$Eh = E^0 + 0.059 \lg K - 0.059 \lg a_{Cl'} \qquad (8)$$

It is not difficult to see that this oxidation—reduction potential reflects the reaction

$$2Hg + 2Cl = Hg_2Cl_2 + 2e$$

From Eq. (8) it is clear that if we add a solution containing chlorine ions to the system $Hg + Hg_2Cl_2$ this will lead to a decline in the oxidation—reduction potential. The processes indicated involve the operating principle of the calomel electrode. The electrode is a glass tube in which a paste of mercury and calomel is placed. This paste is covered by a solution of potassium chloride. If we use a 0.1 N solution of KCl, then at 25°C the potential of this electrode relative to the standard hydrogen electrode has a value E = +0.3369V. If we use a 1 N solution, E = 0.2819 V, and with a saturated solution of KCl, E = 0.2458 V.

Depending on its purpose and function, the second electrode is different. For determining the oxidation–reduction potential a smooth platinum electrode is most frequently used for the second electrode. Gold, iridium, or palladium wire may also be used. It is essential that the metal be noble, i.e., that it possess a large value of the work function for the cation. In this case the effect of the metal on the potential is small. This electrode, connected through a measuring device with the calomel electrode, will merely transfer current, essentially, and, consequently, will characterize the potential difference between an investigated solution and the calomel electrode. If the electrode does not satisfy these requirements, it is then impossible to measure the oxidation–reduction potential. The electrometric determination of ions in solution is based on this principle, For example, replacement of the smooth platinum electrode by a glass electrode permits us to determine the concentration of hydrogen ions in a solution.

The method is basically the following. An electrode is prepared from readily leached glass. Since the resistance of glass is large, the sensitive part of the electrode should have very thin walls (0.01 mm and thinner) in order that the resistance not exceed a few tens of megohms. The inner part of the glass tube, with a thin-walled bulb at the end, is filled with some electrolyte. The electrode is allowed to soak for a long time in an acid solution. Ions of sodium and calcium are leached from the surface of the glass during this period of soaking, and ions of hydrogen replace them. As a result, a gel-like film of swollen $HRSiO_2$ forms on the surface of this electrode. The electrode thus becomes an element "saturated" by cations of hydrogen. Depending on the concentration of hydrogen ions in the

investigated solution, $H^{\cdot}$ ions are removed in some measure from the surface of the electrode, as a result of which the electrode acquires an appropriate charge. According to the potential difference between the calomel and glass electrodes the concentration of $H^{\cdot}$ ions in the solution may be determined.

For determining the activities of sulfide ions we use sulfide–silver electrodes [Kramer and Vail', 1957; Shpeizner and Zaidman, 1965; Berner, 1964]. The potential of the sulfide–silver electrodes is determined by the activity of sulfide ions in the solution.

We have considered only those basic assumptions, applying to thermodynamics, necessary for investigating the conditions under which authigenic minerals form. In doing this, we have tried to present the material in a form that is most important in its application to geological conclusions. The present chapter, therefore, makes no pretension of a detailed discussion of the thermodynamics of oxidation–reduction processes. For our purpose in investigating the principle, many essential aspects of these processes were generally not mentioned. The chapter is designed to help the reader grasp what is meant by equilibrium constant, solubility product, oxidation–reduction potential, and standard potential. It is very important to understand why

$$\lg SP = -\frac{\Delta G^A_{fin} - \Delta G^0_{in}}{1.364},$$

$$Eh = E^0 + \frac{0.059}{n} \lg \frac{oxid.}{reduc.}, \text{ where } E_0 = \frac{\Delta G^A_{fin} - \Delta G^0_{in}}{23.06 \cdot n}$$

since this makes it possible to apply these equations properly.

We should note that these are basic equations, on which the present work is constructed. An understanding of their essential character is therefore necessary. More or less fundamental knowledge of the given problem may be obtained only from a course in thermodynamics.

CHAPTER 3

STABILITY DIAGRAMS OF IRON MINERALS

Stability Boundary of the Natural Environment

Before going on to construction of diagrams, it is advisable that we consider the stability boundaries of the natural environment. In the present work we shall analyze the reactions that take place in an aqueous medium. As a first approximation, therefore, we may say that the limits of formation, existence, and stability of authigenic minerals will be determined by the stability boundaries of water. By knowing the general behavior of thermodynamic plots, it is possible to determine these boundaries. We shall use the following considerations for this purpose. The stability of water from the oxidation-reduction standpoint is limited by two equations:

$$2H_2O = O_2 + 4H^\cdot + 4e \quad E^0 = 1.229$$
$$H_2 = 2H^\cdot + 2e \quad E^0 = 0.$$

From the first equation we find

$$Eh = 1.229 - 0.059pH + 0.015 \lg P_{O_2}$$

If the atmosphere above the water consists exclusively of oxygen and the total pressure is 1 atm, for this limiting case $\lg P_{O_2} = 0$, and

$$Eh = 1.229 - 0.059pH$$

By assigning certain (arbitrary) values of pH, we determine a series of points through which we may draw a straight line characterizing the stability of water relative to oxygen. If the oxidation-reduction potential of the medium is above this line, the water begins to decompose, with the elimination of oxygen, since P_{O_2} exceeds 1 atm. Consequently, no reaction in which water takes part may occur under these conditions. In a similar way the stability zone of water relative to hydrogen may be limited. Let us recall that for the reaction $H_2 = 2H^\cdot + 2e$ the value of normal potential is zero, just as an element for which this reaction takes place is taken as the element of comparison, which therefore characterizes the basis of computation for all possible oxida-

tion-reduction systems. Therefore, for the reaction in question

$$Eh = -0.059pH - 0.029 \lg P_{H_2}$$

As seen from the equation, the stability boundary of water, apart from pH and Eh, depends also on P_{H_2}. Under ordinary conditions, similar to conditions at the surface, the total pressure is 1 atm. If the pressure of hydrogen in the system exceeds atmospheric pressure, water then begins to decompose with the elimination of gaseous hydrogen. Therefore, $P_{H_2} = 1$ atm reflects the greatest possible value at which water may still exist. But in this case $0.029 \lg P_{H_2} = 0$ and, then, Eh $= -0.059$ pH.

The same is true in relation to the oxygen boundary; we may use a number of pH values to determine the stability boundary of water. These boundaries have been plotted on a graph (Fig. 16). The area between the two lines of stability of water

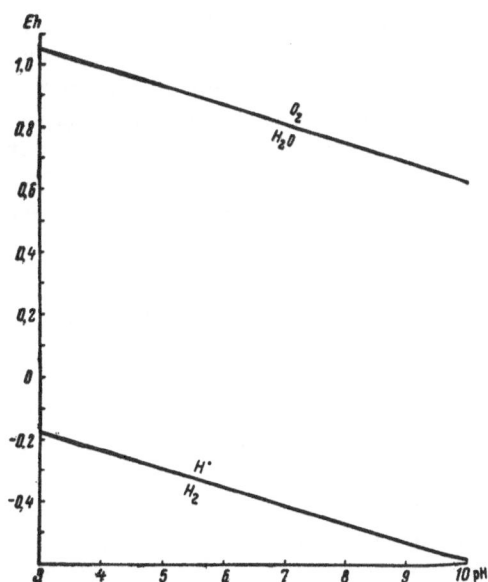

Fig. 16. Boundaries for the existence of aqueous solutions in pH—Eh coordinates.

35

relative to oxygen and hydrogen represents the theoretical stability zone, in which oxidation–reduction reactions in aqueous solutions may take place.

If we go from the thermodynamically possible zone to natural conditions, we find that the actual zones in nature where the indicated reactions may take place is considerably more restricted. This is due to the specific condition which is responsible for creation of the oxidation–reduction situation in nature. In the geologic literature there exist many different and, at times, vague points of view concerning the causes for the origin of a particular oxidation–reduction potential. The process is actually very complex, and perhaps it cannot be unraveled completely at the present time. However, whatever the details that may come to light later, we obtain the proper idea concerning development of oxidation–reduction potential only when the process is examined from two sides. In nature there must exist:

1) Elements that have prepared conditions for the development of some particular oxidation–reduction potential, and

2) Elements that are responsible for establishment of the particular oxidation–reduction potential.

The first includes biological factors, the second chemical.

Biological factors themselves do not determine the oxidation–reduction potential of an environment. They merely produce the preliminary conditions from which a particular oxidation–reduction environment may develop. The actual value of Eh itself depends on other causes. The following examples may help to gain a real understanding of this.

In examining the biochemistry of methane fermentation (a microbiological process), Barker [1936] provided a system of equations to explain the essence of the reduction of sulfates, nitrates, and carbonates:

$$4H_2A + H_2SO_4 \longrightarrow 4A + H_2S + 4H_2O$$
$$4H_2A + HNO_3 \longrightarrow 4A + NH_3 + 3H_2O$$
$$4H_2A + H_2CO_3 \longrightarrow 4A + CH_4 + 3H_2O$$

As a result of these reactions, elements appear that readily give up electrons and are therefore able to participate in the development of a definite oxidation–reduction potential in the environment. Along with this, processes moving in the opposite direction may take place: oxidation of divalent iron to trivalent by bacteria, oxidation of hydrogen sulfide to sulfur by sulfur bacteria, and, lastly, oxidation of reduced elements by the simple

influx of oxygen. In one way or another, if we disregard the reaction products that are forming, microbiological processes are directed at the creation of a kind of electrical potential that is used up in the reduction of certain elements.

What is it that determines directly any particular oxidation–reduction potential? To answer this question we must examine the poise of the medium. From the viewpoint of method, a very good explanation of the poise of a medium is found in the book of Michaélis [1930]. Unfortunately there are several inaccurate treatments of this explanation in the geologic literature [Vainbaum, 1960b]. In the discussion below, we shall follow the explanation of Michaélis, somewhat simplified.

As was shown in the preceding chapter, the oxidation–reduction potential of a medium is determined by the equation

$$Eh = E^0 + \frac{RT}{Fn} \ln \frac{\text{oxidized form}}{\text{reduced form}}$$

Let us designate the oxidized form by y and the total concentration of the oxidation–reduction system by B. Then

$$Eh = E^0 + \frac{RT}{Fn} \ln \frac{y}{B-y}$$

We must call to mind that natural logarithms are used in the indicated equation. We should recall, further, that the base of natural logarithms is e, which is a limit to which the expression $(1 + 1/n)^n$ reduces. If n increases to an infinitely large number, the given expression assumes a perfectly concrete value: 2.7182818... .

Let us add to the investigated system a small amount of a strong oxidizing agent. As a result of this, y increases somewhat, and B − y decreases. If y increases by Δy, then

$$Eh + \Delta Eh = E^0 + \frac{RT}{Fn} \ln \frac{y + \Delta y}{B - (y + \Delta y)}$$

From which we may compute the absolute value of the oxidation–reduction potential:

$$(Eh + \Delta Eh) - Eh = \Delta Eh = E^0 + \frac{RT}{Fn} \ln \frac{y + \Delta y}{B - (y + \Delta y)} -$$
$$- E^0 - \frac{RT}{Fn} \ln \frac{y}{B-y}$$
$$\Delta Eh = \frac{RT}{Fn} \ln \left[1 + \frac{B \, \Delta y}{(B-y) \, y - y \, \Delta y} \right]$$

The degree of change of the oxidation–reduction potential, depending on change in y, will be defined by

$$\frac{\Delta Eh}{\Delta y} = \frac{RT}{Fn} \cdot \frac{1}{\Delta y} \lg \left[1 + \frac{\Delta y}{y \frac{(B-y)}{B} - \frac{y}{B} \Delta y} \right]$$

Let us designate the value

$$\frac{\frac{y(B-y)}{B} - \frac{y}{B} \Delta y}{\Delta y}$$

by n; then

$$\frac{1}{\Delta y} = \frac{n}{\frac{y(B-y)}{B} - \frac{y}{B} \Delta y}$$

Hence

$$\frac{\Delta Eh}{\Delta y} = \frac{RT}{Fn} \cdot \frac{B}{y(B-y) - y \Delta y} \cdot \ln \left(1 + \frac{1}{n} \right)^n$$

If the amount of oxidizing agent added is infinitesimally small, then, instead of Δy we shall have the differential dy, dEh will replace ΔEh, $y dy \to 0$, and $n \to \infty$. Therefore $\ln \left(1 + \frac{1}{n} \right)^n =$ $\ln e = 1$.

The expression for degree of change in oxidation−reduction potential with change in concentration of the oxidized form then takes on the form

$$\frac{dEh}{dy} = \frac{RT}{Fn} \cdot \frac{B}{y(B-y)}$$

The degree of change in oxidation−reduction potential with change in y depends on two variables (B and y), i.e., on the total activity of the participating components and the activity of the oxidized form. Since the value of y is a part of B, if B is taken as unity, y will range from zero to unity. As seen from this equation, the nearer y is to zero or unity, the greater the change in the oxidation−reduction potential. This means that at low amounts of the oxidized form as compared with the reduced form (or at low amounts of the reduced form as compared with the oxidized form) it is sufficient to add a small amount of the oxidized form (or the reduced form in the opposite case) to shift the oxidation−reduction potential markedly. On the other hand, when y = 0.5B, dEh/dy = $(RT/F_n) \cdot 2$, i.e., the change in Eh with small change in y will be negligible in this region. In other words, Eh responds but weakly to small changes in y when the oxidized form and reduced form are present in equal amounts. Michaélis called the value dEh/dy the flexibility of the oxidation−reduction system. In practice it is more convenient to use, not the concept of flexibility, but the concept of poise of the oxidation−reduction system. The poise is the reciprocal of flexibility. Use of poise is more convenient in the sense that this concept approaches the concept of buffering, applicable to pH. Thus,

the poise (β) of the oxidation−reduction system is determined from the equation

$$\beta = \frac{F}{RT} \cdot n \frac{(B-y)}{B} \cdot y$$

As seen from the equation, the poise is characterized by the following parameters: n, B, and y. With respect to n, it may be said that the poise of the oxidation−reduction system increases (and flexibility declines correspondingly) with increase in number of valence electrons participating in the reaction. The effects of the other two variables may be seen from Fig. 17, which helps us to understand that the poise of a system increases with increase in concentration of the participating components (B). The poise of a system also changes with the ratio of oxidized to reduced form. The maximum poise, or stability, of the oxidation−reduction state for a given pair corresponds to the case of approximately equal concentrations of the oxidized and reduced forms and the dominance of this pair over all others in the system. If, for example, sulfates are being reduced in a mud and the interstitial water is characterized by a high content of Fe'' and Fe''', the concentrations of which differ but little, it is the system Fe''' → Fe'' that determines the oxidation−reduction potential. This will continue so long as hydrogen sulfide in the system does not excessively increase the difference in concentration of ferric and ferrous ions. The potential-limiting system will always be the one that has the greatest poise relative to the others.

Biological factors, in favoring reduction processes, thus create conditions for establishing a definite oxidation−reduction potential, but, regardless of the element that is reduced, the oxidation−reduction potential of the medium is determined by the most highly charged system, which in this case is called the potential-limiting system. This naturally does not exclude the possibility that the potential-limiting system is the one on which the microorganisms are acting, but, in general, this condition is not essential.

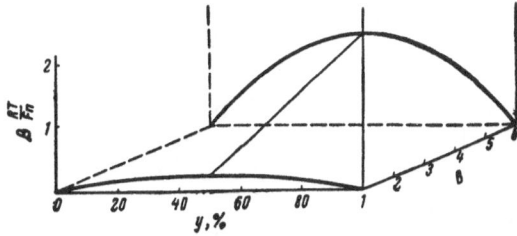

Fig. 17. Change in poise of an oxidation−reduction system with change in reduced form (y) and total concentration of the oxidation−reduction pair (B).

TABLE 4. Parameter Limits of Biological Media within
Their Variation Range for Different Organisms

Organisms	Limits of pH	Range of pH	Limits of Eh, mv	Range of Eh
Algae	1.20—11.75 (12.60)	10.52 (11.40)	+630—220	850
Sulfate-reducing bacteria	4.15—9.92	5.77	+115—450	565
Purple bacteria	4.92—9.75	4.83	+328—230	558
Sulfur bacteria	1.00—9.20	8.20	+855—190	1045
Green bacteria	6.15—9.78	3.63	+7—293	300
Iron bacteria	2.00—8.90	6.90	+850+60	790
Denitrifying bacteria	6.20—10.20	4.00	+665—205	870

In this connection, in studying the effect of a biological system on the oxidation–reduction potential of a medium, it is important to know not only what the given biological group reduces but also what the poise may be of other systems participating in the experiment. Unfortunately, work carried on in this field is vague. We are therefore forced to use absolute data of pH and Eh measurements, which "create" biological factors independent of the conditions in the environment.

A list of such data (Table 4) is found in the basic summary of Baas Becking, Kaplan, and Moore [1960].

After studying a large number of data, these investigators marked out the boundaries of natural (aqueous) media (Fig. 18). As seen from Fig. 18, the stability field is somewhat smaller than the theoretically computed field, and lies completely within it.

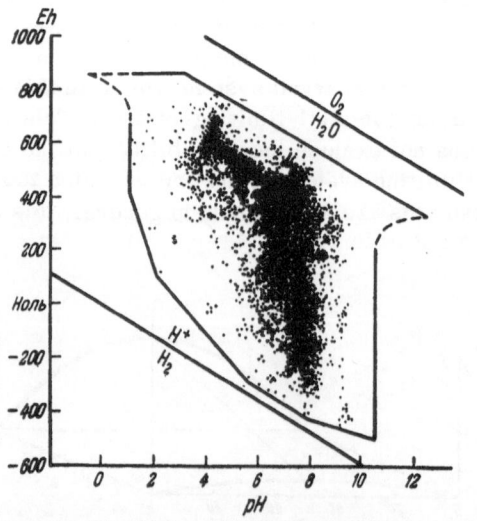

Fig. 18. Distribution of Eh and pH values in natural aqueous media (after Baas Becking, Kaplan, and Moore [1960]).

The System $Fe(OH)_3 - Fe_3O_4 - Fe(OH)_2$

After gaining on acquaintance with the concepts of oxidation–reduction potential, equilibrium constant, and solubility constant, it is advisable to consider the stability conditions of some simple minerals. It is most convenient to begin our examination with magnetite.

Magnetite is a very rare mineral in marine sedimentary rocks. It has been said fairly recently that the mineral may form only at high temperatures [Latimer, 1952]. Geologists have frequently found magnetite, however, that clearly formed under supergene conditions, i.e., at normal temperatures and pressures [Kalganov, 1942; Yanitskii, 1942; Taldykin, 1947; Ginzburg and Rukavishnikova, 1951; Rakhmanov, 1958; Strakhov, 1960; Litvinenko and Drozdov, 1962; and others]. Magnetite was obtained experimentally by Lapteva [1958] and Pavlov [1964] from cold solutions, and this naturally confirms the idea of its possible authigenic formation. Shcherbina also synthesized magnetite under the same conditions. More recently authigenic magnetite has been found in recent lagoonal sediments [Stashchuk et al., 1964]. To explain the conditions of formation of this mineral we may use several reactions and obtain, correspondingly, different fields, in particular

$$Fe_3O_4 + 5H_2O = 3Fe(OH)_3 + H^{.} + e \qquad (9)$$

$$3Fe^{..} + 4H_2O = Fe_3O_4 + 8H^{.} + 2e \qquad (10)$$

$$Fe_3O_4 + 8H^{.} = 3Fe^{...} + 4H_2O + e \qquad (11)$$

In her computations, Lapteva [1958] started with the equation

$$Fe_3O_4 + OH^{.} + 4H_2O = 3Fe(OH)_3 + e.$$

The simplest solution is naturally given by Eq. (9), since solid compounds, water, and the H[.] ion take part in it. The stability field computed from this equation will be therefore determined entirely by the values of Eh and pH. In the other equations, apart from these parameters, the activity of $Fe^{..}$ (10) or $Fe^{...}$ (11) also affects the stability of magnetite. This basis for using Eq. (9) is insufficient, however. The equation is more convenient to use, but it does not reflect the actual conditions. To consider the actual possibilities it is necessary to acquaint ourselves more thoroughly with the behavior of iron ions in solutions of different acidities.

Salts of weak bases, which the salts of iron are, interact in aqueous solutions with the $(OH)'$ ions, with the formation of hydrates. The result of this interaction is called hydrolysis. It is possible to calculate what the activity of $Fe^{..}$ must be in a solution for $Fe(OH)_2$ to be stable in the sediment, or, similarly, to compute the activity of $Fe^{...}$ for the existence of $Fe(OH)_3$ in equilibrium in the sediment. One such method of calculation for $Fe(OH)_3$ is cited in Chap. 2, where it was explained that, for the existence of $Fe(OH)_3$ in the sediment, the relations in the solution must be

$$\lg a_{Fe^{...}} = 2.65 - 3pH$$

We may compute the stability of $Fe(OH)_2$ in exactly the same way:

$$Fe(OH)_2 = Fe^{..} + 2(OH)'$$

In arranging the values of free energies (see Appendix), we found for this reaction

$$\Delta G^0_{reaction} = 2\,\Delta G^0_{(OH)'} + \Delta G^0_{Fe^{..}} - \Delta G^0_{Fe(OH)_2} =$$
$$= -75.19 - 20.30 + 115.57 = 20.26$$

$$\lg K = -\frac{20.26}{1.364} = -14\,853, \quad K = 1.4 \cdot 10^{-15}$$

$$a_{Fe^{..}} \cdot a^2_{(OH)'} = 1.4 \cdot 10^{-15}$$

$$a_{Fe^{..}} = \frac{1.4 \cdot 10^{-15}}{a^2_{(OH)'}}, \quad a_H \cdot a_{OH} = 1.27 \cdot 10^{-14}$$

$$a^2_{(OH)'} = \frac{1.6 \cdot 10^{-28}}{a^2_H}, \quad a_{Fe^{..}} = 8.75 \cdot 10^{12} \cdot a^2_H$$

$$\lg a_{Fe^{..}} = 12.94 - 2pH \tag{12}$$

It is clear that the hydrolysis of iron salts increases with increase in pH. An increase by one unit diminishes the activity of ferric irons a thousandfold and the activity of ferrous ions a hundredfold.

At pH < 3 hydrolysis of ferric ions does not take place, since, for this to occur, there would have to be an improbably high concentration (activity) of this ion in the solution. For ferrous iron, complete control of $Fe^{..}$ as a result of precipitation of $Fe(OH)_2$ begins at a pH of about 6. It becomes understandable, therefore, that below pH 3 the oxidation−reduction state of ferric and ferrous ions does not depend on the pH value but is determined by the system

$$Fe^{..} \rightleftarrows Fe^{...} + e \tag{13}$$

Within the limits of pH of 3 and 6, $Fe(OH)_3$ completely controls the activity of the ferric ion in solution, and the oxidation−reduction system is determined by the equation

$$Fe^{..} + 3H_2O = Fe(OH)_3 + 3H^. + e \tag{14}$$

And, lastly, at pH $\geq$ 6, the oxidation−reduction state of iron is determined by the equation

$$Fe(OH)_2 + H_2O = Fe(OH)_3 + H^. + e \tag{15}$$

These data are sufficient for us to speak of the applicability of Eq. (11) for very acidic conditions, Eq. (10) for acid conditions, and Eq. (9) for alkaline conditions. Precise boundaries of applicability (range of pH) of the indicated equations may be determined after we investigate the stability boundaries of $Fe^{..}$ and $Fe(OH)_3$ in pH−Eh coordinates.

Leaving Eq. (13) and, consequently, the reaction of (11) without further attention, since we are considering conditions for pH $\geq$ 3, we may determine the conditions for occurrence of the different forms of iron from Eqs. (14) and (15).

The Appendix contains values of free energies used in such computations. Some of these values, particularly those for iron hydroxide, magnetite, and the ferric ion, were obtained by computation. What necessity led us to depart, for these materials, from data given in handbooks? The value of standard free energy of magnetite given by Latimer [1952] is −242.400 kcal. This applies to well-crystallized material. The free energy of magnetite of the type $Fe_3O_4 \cdot xH_2O$, which may be obtained from cold solutions by reduction of iron hydroxide, has not been measured. The existence of this compound is indicated in his thermodynamic tables by Pourbaix [1963], but no data on the free energy are given. This investigator therefore includes magnetite in the diagram in which the initial material is Fe_2O_3. In the diagrams plotted on the basis of ferric hydroxide, Fe_3O_4 is absent. In 1958 Lapteva obtained magnetite from cold solutions during experimental work on ferric hydroxide. The following data were determined during this experiment

$$Fe^{..} \rightleftarrows Fe^{...} + e, \quad E^0 = 0.738$$

$$Fe^{..} + 3H_2O \rightleftarrows Fe(OH)_3 + 3H^. + e, \quad E^0 = 0.908$$

$$3Fe^{..} + 4H_2O \rightleftarrows Fe_3O_4 + 8H^. + 2e, \quad E^0 = 1.206$$

$$\lg a_{Fe^{..}} + 3 \lg a_{(OH)'} = -39.43$$

In regard to reaction (13) we should note that the potential 0.738V, taken from Lapteva, characterizes the lowest value of this system. The highest value of the standard potential ascribed to this system is 0.783 V. In the recent literature we most commonly find the value 0.77 V [Latimer, 1952; Pourbaix, 1963]. Most investigators who have determined E^0 experimentally for the ferric—ferrous system give the figure 0.745 V [Lapteva, 1958]. In view of the fact that the reliability of the standard potential for the system (13) within the range 0.735 to 0.77V is practically invariant, we adopt the value $E^0 = 0.75$ V. Thus, for determining standard free energies, we have assembled the following data, obtained by Lapteva with our correction in system (13):

$$Fe^{\cdot\cdot} \rightleftarrows Fe^{\cdot\cdot\cdot} + e, \quad E^0 = 0.75, \quad \Delta G^0 = 17.30 \tag{13}$$

$$Fe^{\cdot\cdot} + 3H_2O \rightleftarrows Fe(OH)_3 + 3H^{\cdot} + e,$$
$$E^0 = 0.908, \quad \Delta G^0 = 20.94 \tag{14}$$

$$3Fe^{\cdot\cdot} + 4H_2O \rightleftarrows Fe_3O_4 + 8H^{\cdot} + 2e,$$
$$E^0 = 1.206, \quad \Delta G^0 = 55.62 \tag{10}$$

$$Fe^{\cdot\cdot\cdot} + 3(OH)^{'} = Fe(OH)_3, \quad \Delta G^0 = 53.78 \tag{16}$$

On the basis of reactions (13), (14), and (16), we may set up the following system:

$$(I) \begin{cases} \Delta G^0_{Fe(OH)_3} - \Delta G^0_{Fe^{\cdot\cdot\cdot}} = -166.56 \\ \Delta G^0_{Fe(OH)_3} - \Delta G^0_{Fe^{\cdot}} = -149.13 \\ \Delta G^0_{Fe^{\cdot\cdot\cdot}} - \Delta G^0_{Fe^{\cdot\cdot}} = -17.30 \end{cases}$$

This system may be solved relative to two variables, which makes it possible to determine the reliability of the data obtained:

$$\Delta G^0_{Fe(OH)_3} - \Delta G^0_{Fe^{\cdot\cdot}} = -149.13$$
$$\Delta G^0_{Fe(OH)_3} - \Delta G^0_{Fe^{\cdot\cdot}} = -149.26$$

The use of $E^0 = 0.738$ V, as done by Lapteva, would greatly increase the divergence. The values obtained show rather good agreement. By using their average values and the data of reaction (10), we form a new system:

$$(II) \begin{cases} \Delta G^0_{Fe(OH)_3} - \Delta G^0_{Fe^{\cdot\cdot}} = -149.19 \\ \Delta G^0_{Fe_3O_4 \cdot xH_2O} - \Delta G^0_{Fe^{\cdot\cdot}} = -171.14 \end{cases}$$

From the three values used in system (II), the greatest reliability is found in the data for magnetite and the ferric ion. The data on magnetite apply to well-crystallized material. However, X-ray analysis of magnetite that has formed in recent lagoonal sediments indicates a weakly crystallized state for this material, at least in the incipient stages of formation [Stashchuk et al., 1964]. It is therefore logical to assume that, for this form of magnetite, the standard free energy will differ somewhat from that adopted for the crystalline state in a way similar to that for other compounds [Levchenko, 1950]. Therefore, in solving the equations of (II), $\Delta G^0_{Fe^{\cdot\cdot}}$ should be assigned a value. According to Latimer [1954], $G^0_{Fe^{\cdot\cdot}} = -20.30$. Whence, from system (II),

$$\Delta G^0_{Fe(OH)_3} = -169.49; \quad \Delta G^0_{Fe_3O_4 \cdot xH_2O} = -232.04$$

On the basis of the value $E^0 = 0.75$, adopted by us for reaction (13), we get, in agreement with the standard free energy of the ferric iron,

$$\Delta G^0_{Fe^{\cdot\cdot\cdot}} = -20.30 + 17.30 = -3.00$$

For reaction (14), according to a model already known, we determine the standard potential

$$\Delta G_{reaction} = \Delta G^0_{Fe(OH)_3} - (3\,\Delta G^0_{H_2O} + \Delta G^0_{Fe^{\cdot\cdot}})$$
$$E^0 = \frac{\Delta G^0_{reaction}}{23.06 \cdot n}$$

After substitution of the appropriate data from the Appendix, we find

$$Eh = 0.907 + 0.059 \lg \frac{a_{Fe(OH)_3} \cdot a^3_{H^{\cdot}}}{a_{Fe^{\cdot\cdot}} \cdot a^3_{H_2O}}.$$

Since, for dilute solutions,

$$\lg a_{Fe(OH)_3} = \lg a_{H_2O} = 1$$

then

$$Eh = 0.907 + 0.059 \lg \frac{a^3_{H^{\cdot}}}{a_{Fe^{\cdot\cdot}}}$$

and

$$Eh = 0.907 - 0.177 pH - 0.059 \lg a_{Fe^{\cdot\cdot}}. \tag{17}$$

In a similar way we may find that the equilibrium line of ferric hydroxide—ferrous hydroxide (15) is determined by the equation

$$Eh = 0.12 - 0.059 pH.$$

The last equation may be shown very well in Eh—pH coordinates, but the equilibrium of reaction (14) is determined, apart from Eh and pH, also by the activity of the ferrous ion. When plotting data on graphs for these two equations, it is desirable to use a three-dimensional diagram, on which different values of activity of the ferrous ion are placed on the G axis.

As seen from Fig. 19, to the left of the plane of $Fe(OH)_3 - Fe^{\cdot\cdot}$ the ferrous iron is not controlled by ferrous hydroxide, whereas to the right of this

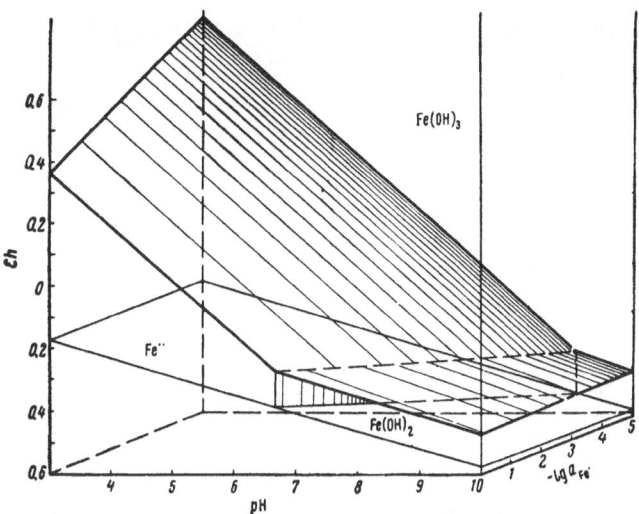

Fig. 19. Intermediate diagram for occurrence of $Fe^{\cdot\cdot}$, $Fe(OH)_3$, and
$Fe(OH)_2$.

plane we have to do with $Fe^{\cdot\cdot\cdot}$, the content of which
is strictly determined by the solubility of ferric
hydroxide in the sediment. These data completely
determine the possible variants in the use of equations for investigating the stability fields of magnetite. In particular, reaction (9) may be used within
the limits of the field where $Fe(OH)_3$ may exist. It
is true that it would formally be more proper to use
the reaction proposed by Lapteva, since it is stress-
ed in this case that the reaction takes place in an
alkaline environment. But Eq. (9), with fewer transformations, gives the same results when going on to
the Nernst equation. We should remember, merely,
that this equation is applicable only in the field

where $Fe(OH)_3$ exists. In going over to the field of
$Fe^{\cdot\cdot}$, it is necessary to use reaction (10).

Disposing of the investigated boundaries, we
may go on to determination of the stability fields of
magnetite.

For alkaline conditions, using the general
methods of computation in reaction (9), the boundary
plane of magnetite—ferric hydroxide is determined
by the expression

$$Eh = 0.31 - 0.059 pH \qquad (18)$$

This expression is the equation of a straight
line. For plotting on the graph it is therefore sufficient to assign two values of pH. In similar fashion,

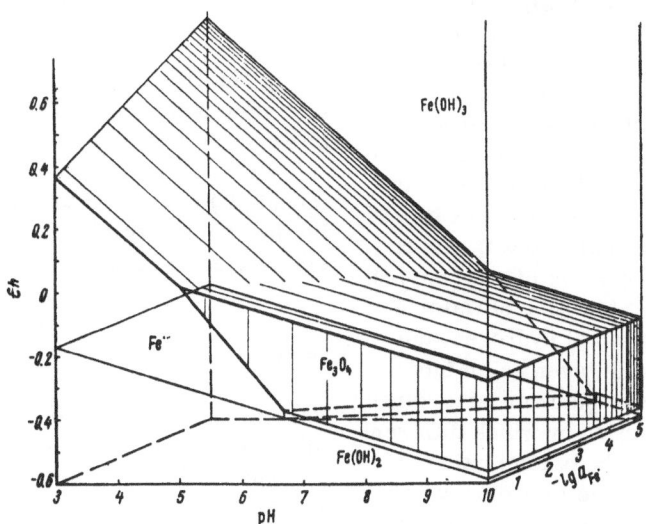

Fig. 20. Stability zones of $Fe^{\cdot\cdot}$, $Fe(OH)_3$, Fe_3O_4, and $Fe(OH)_2$.

for acidic conditions, we obtain for Eq. (10)

$$Eh = 1.206 - 0.236pH - 0.089 \lg a_{Fe^{..}}$$

We have plotted a graph from these data, showing the stability field of magnetite in the Eh–pH system (Fig. 20).

As may be seen from a comparison of Figs. 19 and 20, the magnetite zone covers the zone of ferrous hydroxide. This makes it necessary to re-examine the equilibrium plane $Fe(OH)_3 - Fe(OH)_2$ and to replace it by the equilibrium $Fe_3O_4 - Fe(OH)_2$. For investigating the stability boundary relative to $Fe(OH)_2$, we shall start from the reaction

$$3Fe(OH)_2 = Fe_3O_4 + 2H_2O + 2H^. + 2e$$
$$Eh = 0.03 - 0.059pH$$

For the standard potential it is seen that the upper boundary of ferrous hydroxide is shifted appreciably toward the reducing zone and is found near the stability boundary of water.

After doing the necessary plotting, we shall now analyze the graphical material (Fig. 20). The planes obtained on the graph divide the equilibrium zones of occurrence of the solid phases in the sediment according to the oxidation–reduction potential and pH of the environment. In the given system there can be no occasion when ferric hydroxide, ferrous hydroxide, and magnetite may coexist simultaneously in equilibrium. Only pairs of the three compounds may coexist in equilibrium: either $Fe(OH)_3$ and Fe_3O_4, or Fe_3O_4 and $Fe(OH)_2$. On the other hand, the coexistence of $Fe(OH)_3$ and $Fe(OH)_2$ is excluded.

The following special features should be noted. Since, in normal natural processes, the activity of the ferrous ion is always considerably below 1 mole per liter (i.e., somewhat greater than 55.85 g/liter depending on ionic force and degree of ionization), magnetite forms under rather restricted alkaline conditions. With decline in the oxidation–reduction potential and diminished activity of the ferrous ion, the magnetite field shifts sharply toward the alkaline zone, thereby limiting the possibility of its formation in ordinary sedimentary rocks. Apart from this, since the boundary of the natural environment passes somewhat above the zone of water decomposition with the separation of hydrogen, the formation of ferrous hydroxide, even as an intermediate product, does not take place under ordinary natural conditions. The reaction by which different compounds of iron form with the participation of $Fe(OH)_2$ is therefore inapplicable to ordinary natural conditions, and we should assign it to the class of doubtful possibilities.

The System $Fe(OH)_3 - Fe_3O_4 - Fe(OH)_2 + CO_2$

The possibility of siderite forming may be determined by the reaction

$$FeCO_3 = Fe^{..} + CO_3^{..}$$

To investigate the stability field of siderite, it is necessary to gather information on $a_{Fe^{...}}$, Eh relating $a_{Fe^{...}}$ and the equilibrium $Fe^{...}$ ions, and pH, as well as data on the activity of $CO_3^{..}$, the value of which is defined by pH and P_{CO_2} or ΣCO_2 but is not related directly to Eh. Thus, the minimal number of independent parameters defining the stability of siderite is four. The construction of a three-dimensional diagram is therefore possible only when one of the parameters is kept constant. In addition to this complication, there is still another, associated with the inconvenient use of $a_{CO_3^{..}}$ as one of the parameters. The activity of $CO_3^{..}$ itself depends on the partial pressure of CO_2 or the total dissolved carbon dioxide and the pH. From the formal point of view, both the value of ΣCO_2 and that of P_{CO_2} are rather commonly used parameters in the geologic literature. It would seem, on principle, to be a matter of indifference which one of these is used to investigate the stability regions of minerals. As we shall show, however, diagrams obtained by using P_{CO_2} and ΣCO_2 differ appreciably. This follows from the fact that P_{CO_2} and ΣCO_2 do not have a rectilinear correlation throughout the range of pH.

As is known, $a_{CO_3^{..}}$ is related to CO_2 by the following:

$$CO_2 + H_2O \rightleftarrows H_2CO_3$$
$$H_2CO_3 \rightleftarrows H^. + HCO_3'$$
$$HCO_3' \rightleftarrows H^. + CO_3^{..}$$

The transition $CO_2 + H_2O \rightarrow H_2CO_3$ takes place to a very insignificant degree. Almost always carbon dioxide is found as molecularly dissolved gas in water, and only in very insignificant amounts (about 1%) does it go to carbonic acid. In view of the low solubility of carbon dioxide in water, the equilibrium constant

$$\frac{H_2CO_3}{CO_2 \cdot H_2O} = K$$

is very small.

In our further discussions it will not be necessary to use this constant, but it is generally necessary to keep in mind the fact that it exists. The carbonic acid that forms dissociates according to the scheme

$$H_2CO_3 \rightleftarrows H^. + HCO_3'$$

with the true dissociation constant

$$\frac{a_{H^{\cdot}} \cdot a_{HCO_3'}}{a_{H_2CO_3}} = K_1'$$

Expressing the denominator by $H_2O + CO_2$, we may write that

$$\frac{a_{H^{\cdot}} \cdot a_{HCO_3'}}{a_{CO_2} \cdot H_2O \cdot K} = K_1'$$

Since the reaction takes place in an aqueous medium, the activity of H_2O may be taken as unity, and the expression we obtain takes the following form:

$$\frac{a_{H^{\cdot}} \cdot a_{HCO_3'}}{a_{CO_2 \text{ (dis. in water)}}} = K \cdot K_1' = K_1 \qquad (19)$$

The value $K \cdot K_1' = K_1$ may be rather easily determined experimentally. It is called the first apparent dissociation constant (in contrast to K, which is the true dissociation constant).

The second apparent dissociation constant is

$$\frac{a_{H^{\cdot}} \cdot a_{CO_3''}}{a_{HCO_3'}} = K_2 \qquad (20)$$

since $a_{HCO_3'}$ is related to $a_{H_2CO_3}$ and the latter to CO_2 by the true equilibrium constant K, which applies without limit under our conditions.

The convenience of using apparent equilibrium constants is that, first, we may recompute $a_{H_2CO_3}$, $a_{HCO_3'}$, and $a_{CO_3''}$ to total dissolved carbon dioxide and not to that which proves to be dissociated, and, second, Henry's law, concerning the solubility of a gas in water, applies with no particular error.

Thus, if we use apparent dissociation constants, we may then write that

$$\sum CO_2 = H_2CO_3 + HCO_3' + CO_3'$$

where $\sum CO_2$ is dissolved carbon dioxide.

The relations between the partial pressure and the total dissolved carbon dioxide may be brought out in the following manner:

$$H_2O + CO_{2(r)} = H_2CO_3, \quad \Delta G^0 = 1.9498, \quad K = -1.429$$

$$\frac{a_{H_2CO_3}}{P_{CO_2}} = 10^{-1.429}, \quad a_{H_2CO_3} = 10^{-1.429} \cdot P_{CO_2}$$

Using the dissociation scheme of carbonic acid

$$H_2CO_3 = H^{\cdot} + HCO_3', \quad \frac{a_{H^{\cdot}} \cdot a_{HCO_3'}}{a_{H_2CO_3}} = K_1 = 10^{-6.371}$$

$$HCO_3' = H^{\cdot} + CO_3'', \quad \frac{a_{H^{\cdot}} \cdot a_{CO_3''}}{a_{HCO_3'}} = K_2 = 10^{-10.33}$$

we determine, correspondingly,

$$a_{HCO_3'} = \frac{10^{-7.8}}{a_{H^{\cdot}}} \cdot P_{CO_2}$$

$$a_{CO_3''} = \frac{10^{-18.13}}{a_{H^{\cdot}}^2} \cdot P_{CO_2} \qquad (21)$$

On this basis we find

$$\sum CO_2 = H_2CO_3 + HCO_3' + CO_3'' =$$
$$= P_{CO_2}\left(10^{-1.429} + \frac{10^{-7.8}}{a_{H^{\cdot}}} + \frac{10^{-18.13}}{a_{H^{\cdot}}^2}\right) \qquad (22)$$

Careful examination of the sum shown in parentheses shows that the terms are not equivalent. The significance of each term depends on the pH value. It is not hard to ascertain that the first term will be equal to the second at a pH of 6.37, since

$$10^{-1.429} = \frac{10^{-7.8}}{a_{H^{\cdot}}}, \quad a_{H^{\cdot}} = 10^{-6.37}, \quad pH = 6.37$$

This means that when pH < 6.37 the second term may be neglected, because its value in this case is less than $10^{-1.429}$. Even more is it possible to neglect the third term, since its significance in comparison with the first two begins at pH > 10.33. The relations between $\sum CO_2$ and P_{CO_2} in explicit form may be therefore represented by the following three particular equations:

$$pH \leqslant 6.37 \quad \sum CO_2 = P_{CO_2} \cdot 10^{-1.429} = 0.037 \cdot P_{CO_2} \quad (23)$$

$$6.37 < pH < 10.33 \quad \sum CO_2 = P_{CO_2} \cdot \frac{10^{-7.5}}{a_{H^{\cdot}}} \quad (24)$$

$$pH > 10.33 \quad \sum CO_2 = P_{CO_2} \cdot \frac{10^{-18.13}}{a_{H^{\cdot}}^2} \quad (25)$$

A graph plotted from these values for $P_{CO_2} = 1$ atm is shown in Fig. 21.

As these data show, beginning with a pH of 6.37 the total CO_2 increases with increase in pH despite the fact that P_{CO_2} remains constant. This takes place because of the high solubility of the

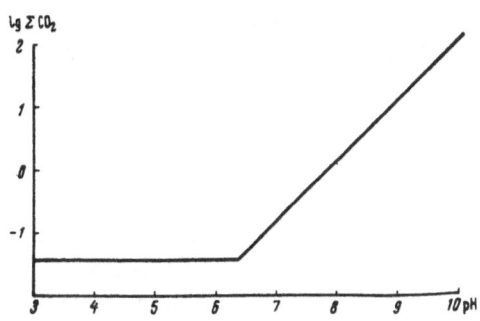

Fig. 21. Relations of log $\sum CO_2$ and pH for $P_{CO_2} = 1$ atm above the solution.

CO_3'' ion with increase in pH. Thus, diagrams plotted by using P_{CO_2} = const reflect a sequential and uniform increase of the CO_3'' ion with increasing alkalinity [see Eq. (21)], but diagrams plotted on the principle that ΣCO_2 = const describe a more complex function of change of CO_3'' with increasing alkalinity and its constancy under alkaline conditions.

Thus, before we begin to plot diagrams in which the character of mineral changes in the investigated system should be reflected, it is necessary to make clear:

1. what parameter to use when plotting the diagram (P_{CO_2} or ΣCO_2?), and
2. what parameter must be taken as a constant (pH, Eh, $a_{Fe\cdot\cdot}$, P_{CO_2}, or ΣCO_2?).

Garrels suggests that diagrams plotted by using P_{CO_2} = const be considered diagrams characterizing open systems, since it is possible to have constant P_{CO_2} by continuous passage of air with a given P_{CO_2} through a system in which pH changes arbitrarily. Diagrams prepared by using ΣCO_2 = const must be assigned to closed systems, in which, with change in pH, only the relations of ions of H_2CO_3, HCO_3', and CO_3'' with invariant ΣCO_2 may change. Unfortunately, neither diagram is hardly ever applicable in its ideal form to natural processes. We must always consider some error introduced by the actual conditions. The most suitable material, particularly for using diagrams plotted according to the principle of ΣCO_2 = const, is a carbonate-free weathering crust, in which an appreciable jump in pH occurs with vertical movement of water. In a weathering crust containing carbonate, diagrams plotted according to the principle P_{CO_2} = const are more applicable. For investigating the formation of minerals in the sea, neither diagram is very suitable, as we shall point out in detail.

Taking into account the specific features of applicability of diagrams to different conditions, let us turn now to the technique of plotting them.

The solubility product for the reaction $Fe\cdot\cdot + CO_3'' = FeCO_3$ is found in the ordinary way:

$$\Delta G^0 = 14.54 \text{ kcal}$$

$$\Delta G^0 = -RT \ln K = -1.364 \lg K, \quad K = 2.2 \cdot 10^{-11}$$

Thus $a_{Fe\cdot\cdot} \cdot a_{CO_3''} = 2.2 \cdot 10^{-11}$.

In the further development it will be more convenient for us to use the logarithmic form:

$$\lg a_{Fe\cdot\cdot} + \lg a_{CO_3''} = -10.6598 \qquad (26)$$

For reaction (14) we have already computed the activity of the ferrous ion in solution relative to the solid phase of ferric hydroxide according to values of Eh and pH (for pH values greater than 3). As computations have shown, the straight line representing equilibrium of the ferrous ion with the solid phase (ferric hydroxide) is determined by Eq. (17). On the other hand, if the activity of the ferrous ion is known, the activity of CO_3'' sufficient for satisfying the solubility product and the precipitation of siderite is fully determined by Eq. (26). Therefore, after substituting the value of the ferrous ion from Eq. (17) in Eq. (26), we may determine the conditions of siderite occurrence according to Eh, pH, and $a_{CO_3''}$:

$$Eh = 1.536 - 0.177 pH + 0.059 \lg a_{CO_3''} \qquad (27)$$

Since the possibility of siderite occurring, according to Eq. (27), is determined by three parameters, the equilibrium of siderite–ferric hydroxide in the field of occurrence of solid ferric hydroxide will be characterized not by an equilibrium line but by an equilibrium surface, plotted in three dimensions with the coordinates Eh, pH, and $a_{CO_3''}$. Until the dependence of $a_{CO_3''}$ on ΣCO_2 and P_{CO_2} is determined, we cannot say what form this surface will have. Using Eqs. (19) and (20), we find

$$\Sigma CO_2 = a_{CO_3''} \left(\frac{a_{H\cdot}^2}{K_1 \cdot K_2} + \frac{a_{H\cdot}}{K_2} + 1 \right)$$

whence

$$a_{CO_3''} = \frac{\Sigma CO_2}{\left(\frac{a_{H\cdot}^2}{K_1 \cdot K_2} + \frac{a_{H\cdot}}{K_2} + 1 \right)} \qquad (28)$$

The expression obtained may be substituted in Eq. (27) also to find the conditions for formation of siderite according to Eh, pH, and ΣCO_2:

$$Eh = 1.536 - 0.177 pH +$$

$$+ 0.059 \lg \Sigma CO_2 - 0.059 \lg \left(\frac{a_{H\cdot}^2}{K_1 \cdot K_2} + \frac{a_{H\cdot}}{K_2} + 1 \right) \quad (29)$$

Equation (29) is sufficient for plotting diagrams of siderite stability in equilibrium with ferric hydroxide under the condition ΣCO_2 = const, but in

---

†This equation might have been obtained in a shorter way, specifically if we start from the reaction $FeCO_3 + 3H_2O = Fe(OH)_3 + 3H^+ + CO_3'' + e$. For greater clarity we have arrived at Eq. (27) by stages. But, on the other hand, this reaction more graphically demonstrates the fact that Eq. (27) is applicable only to the field of ferric hydroxide.

practical use it is somewhat inconvenient because the last member has the logarithm of a sum. Before using it, therefore, it would be desirable to determine the significance of each term of the sum according to the pH value. For this purpose we make a termwise comparison:

$$\frac{a_{H^.}^2}{K_1 \cdot K_2} = \frac{a_{H^.}}{K_2}, \quad a_{H^.} = K_1$$

This means that the two investigated terms are equivalent when pH = $-\lg K_1$. When pH < $-\lg K_1$ the first member will be much greater than the second, and, consequently, the second member may be neglected. When pH > $-\lg K_1$, correspondingly, the second term will be greater than the first, and the first may be therefore neglected. In comparing the second and third terms, we find

$$\frac{a_{H^.}}{K_2} = 1, \quad a_{H^.} = K_2$$

The condition of equality reflects the case when pH = $-\lg K_2$. A convenient form for the last member of Eq. (29) may therefore be represented by three particular cases:

$$pH < -\lg K_1 = 6.37$$

$$\lg\left(\frac{a_{H^.}^2}{K_1 \cdot K_2} + \frac{a_{H^.}}{K_2} + 1\right) \approx \lg\frac{a_{H^.}^2}{K_1 \cdot K_2} = -2pH + 16.7$$

$$6.37 < pH < 10.33$$

$$\lg\left(\frac{a_{H^.}^2}{K_1 \cdot K_2} + \frac{a_{H^.}}{K_2} + 1\right) \approx \lg\frac{a_{H^.}}{K_2} = -pH + 10.33$$

$$pH > -\lg K_2 = 10.33$$

$$\lg\left(\frac{a_{H^.}^2}{K_1 \cdot K_2} + \frac{a_{H^.}}{K_2} + 1\right) \approx \lg 1 = 0$$

Accordingly, we obtain three equations for different pH ranges, characterizing the conditions for siderite formation according to pH, Eh, and ΣCO_2:

$$pH < 6.37, \quad Eh = 0.55 - 0.059pH + 0.059\lg\sum CO_2 \quad (30)$$

$$6.37 < pH < 10.33, \quad Eh = 0.93$$

$$-0.118pH + 0.059\lg\sum CO_2 \quad (31)$$

$$pH > 10.33, \quad Eh = 1.54 - 0.177pH + 0.059\lg\sum CO_2 \quad (32)$$

Equations (30)−(32) are used to investigate equilibrium where the initial material is $Fe(OH)_3$, but this alone specifies the existence of the system $Fe(OH)_3 - Fe_3O_4 - Fe(OH)_2$, which we have already analyzed. Investigation of the stability zone of sid-

erite is therefore possible only by considering this system. It is most convenient under the circumstances to use a graphical method, superimposing each time the planes that characterize the particular equilibria with siderite on the graph (see Fig. 20). But it is impossible to plot values with different ΣCO_2 on a single graph where three coordinate axes are already being used. The simplest solution to this problem, without great loss of clarity, is to prepare a series of particular graphs, plotted for several selected values of ΣCO_2.

Let us begin, for example, with a value of $\Sigma CO_2 = 10^{-2}$ mole/liter. Equations (30)−(32) define the equilibrium of siderite relative to ferric hydroxide. Equilibrium planes may therefore be drawn only up to the junction with the $Fe(OH)_3 - Fe^{...}$ plane under acidic conditions and to the $Fe(OH)_3 - Fe_3O_4$ plane under alkaline conditions. In penetrating the limits of the stability zones of $Fe^{..}$ or Fe_3O_4, one must compute the equilibrium of siderite relative to these components.

Toward an acidic environment, the plane plotted for Eq. (30) intersects the $Fe(OH)_3 - a_{Fe^{..}}$ plane. The formation of siderite in the $Fe^{..}$ zone does not depend on Eh, since $a_{Fe^{..}}$ is a constant value, determined by the value of the coordinate z. A decisive role in this situation is played by $a_{CO_3''}$, which determines whether the solubility product is satisfied relative to Eq. (26). The activity of the carbonate ion at a fixed ΣCO_2 is determined by the pH value. This furnishes a basis for stating that the zone of $FeCO_3$ occurrence is separated from the $Fe^{..}$ zone by a plane passing parallel to the yz plane.

For computing this plane we have used the following data: $\Sigma CO_2 = 10^{-2}$ mole/liter; $\lg a_{Fe^{..}}$ is assigned a series of values from 0 to −5. When $\lg a_{Fe^{..}} = 0$, then, in agreement with Eq. (26), $\lg a_{CO_3''} = -10.6598$. When $\lg a_{CO_3''} < -10.6598$, siderite will be unstable. The stability zone of siderite will be determined, therefore, by the inequality $\lg a_{CO_3''} \geq -10.6598$. Using Eq. (28), we may specify

$$\lg\sum CO_2 - \lg\left(\frac{a_{H^.}^2}{K_1 \cdot K_2} + \frac{a_{H^.}}{K_2} + 1\right) \geq -10.6598$$

but, since for an acidic environment

$$\lg\left(\frac{a_{H^.}^2}{K_1 \cdot K_2} + \frac{a_{H^.}}{K_2} + 1\right) \approx -2pH + 16.7$$

then $2pH \geq 6.0 - \lg\sum CO_2$ and when $\sum CO_2 = 10^{-2}$, pH = 4.

In similar fashion we may compute that when $\lg a_{Fe^{..}} = -5, pH \geq 6.5$. Intermediate values of $\lg a_{Fe^{..}}$

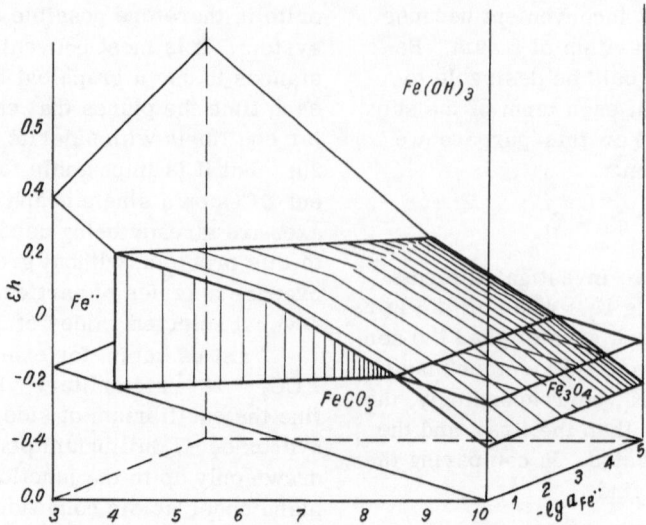

Fig. 22. Stability zones of Fe··, Fe(OH)₃, Fe₃O₄, and FeCO₃ at
$\Sigma CO_2 = 10^{-2}$ mole/liter.

between 0 and −5 have correspondingly intermediate pH values for the formation of siderite.

The conditions under which siderite forms in an alkaline environment may be investigated from the equation

$$3FeCO_3 + 4H_2O = Fe_3O_4 + 3CO_3'' + 8H^· + 2e$$
$$\Delta G^0 = 99.24, \quad E^0 = 2.152$$
$$Eh = 2.152 - 0.236\,pH + 0.088 \lg a_{CO_3''}.$$

Substituting ΣCO_2 for $a_{CO_3''}$, we find

$$Eh = 1.24 - 0.148\,pH + 0.088 \lg \Sigma CO_2$$

Assuming $\Sigma CO_2 = 10^{-2}$, the final computed equation for $Fe_3O_4 - FeCO_3$ paragenesis is

$$Eh = 1.06 - 0.148\,pH.$$

For the data obtained we have plotted a graph of mineral paragenesis in Eh−pH−Fe·· coordinates (Fig. 22). Two interesting details are noteworthy: when $\Sigma CO_2 = 10^{-2}$ mole/liter in weakly acidic, neutral, and weakly alkaline media, magnetite characterizes a more reducing environment than siderite, despite the fact that it contains ferric oxide. Only under strongly alkaline conditions does siderite occupy its legitimate place, below magnetite according to the Eh value. Under acidic conditions, when pH < 4, the formation of siderite is generally excluded, even at the highest values of ΣCO_2; this is possible in nature with sufficiently high values for the activity of the ferrous ion. With decline in this activity the region of siderite occurrence shifts to-

ward alkaline conditions. The same thing occurs with diminution of ΣCO_2. The effect of ΣCO_2 on the graph is not so clear as the effect of the activity of the ferrous ion, but one may convince himself that this is so after an investigation of Eqs. (30)−(32). Actually, in these equations log ΣCO_2 is present with a positive sign, and any decline in ΣCO_2 leads to a corresponding lowering of the equilibrium plane. In this case it is of interest to ask: At what ΣCO_2 will siderite begin to form in all zones, restricted by natural conditions, at an oxidation−reduction potential below that of the zone of magnetite occurrence? As seen from Fig. 22, the planes $Fe(OH)_3 - Fe_3O_4$ and $Fe(OH)_3 - FeCO_3$ are parallel to each other at pH < 6.37. In other words, they change according to the same law.†

Therefore, when ΣCO_2 declines, the instant may arrive at which these planes unite, reflecting identical changes in Eh. Setting Eqs. (18) and (30) equal to each other according to Eh, we find

$$0.059 \lg \Sigma CO_2 = -0.24$$
$$\lg \Sigma CO_2 = -4.068, \quad \Sigma CO_2 = 8.5 \cdot 10^{-5}$$

We must therefore conclude that at a particular pH value, with $\Sigma CO_2 > 8.5 \cdot 10^{-5}$ mole/liter, siderite will form under conditions more strongly oxidizing than magnetite will. With $\Sigma CO_2 < 8.5 \cdot$

---

†This is perfectly clear analytically, since, from Eqs. (18) and (30), $\partial Eh/\partial pH = 0.059$.

10^{-5} siderite begins to form under conditions more strongly reducing than those in which magnetite begins to form.

An interesting example of paragenesis is shown in the particular diagram representing $\Sigma CO_2 = 8.5 \cdot 10^{-5}$. Since the $Fe(OH)_3 - Fe_3O_4$ and $Fe(OH)_3 - FeCO_3$ planes within the limits $3 \le pH < 6.37$ coincide, it is necessary to make supplementary investigation of the stability of magnetite relative to siderite, since it is not clear whether these minerals with coincident planes at the boundaries of the indicated pH values, will form independently from ferric hydroxide, or whether one will prove to be unstable relative to the other. To examine this problem we shall use the general equation of siderite-magnetite paragenesis (Eq. (33)), expressing $a_{CO_3''}$ by ΣCO_2 for the condition pH < 6.37. As a result, we obtain the equation $Eh = 0.674 - 0.059pH + 0.088 \lg \Sigma CO_2$.

When we substitute $\Sigma CO_2 = 8.5 \cdot 10^{-5}$ mole per liter, we find

$$Eh = 0.31 - 0.059pH$$

Thus, under the restricted conditions we have imposed for the plane of (18), magnetite should begin to form, but, according to the results obtained

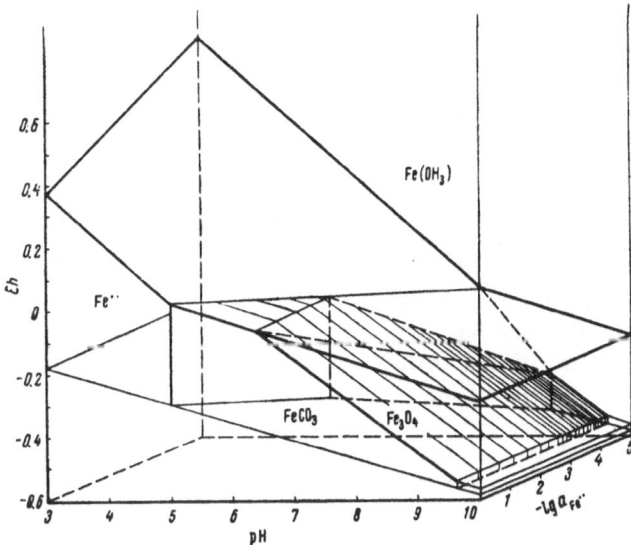

Fig. 23. Stability regions of $Fe^{\cdot\cdot}$, $Fe(OH)_3$, Fe_3O_4, $FeCO_3$, and $Fe(OH)_2$ at $\Sigma CO_2 = 8.5 \, 10^{-5}$ mole/liter.

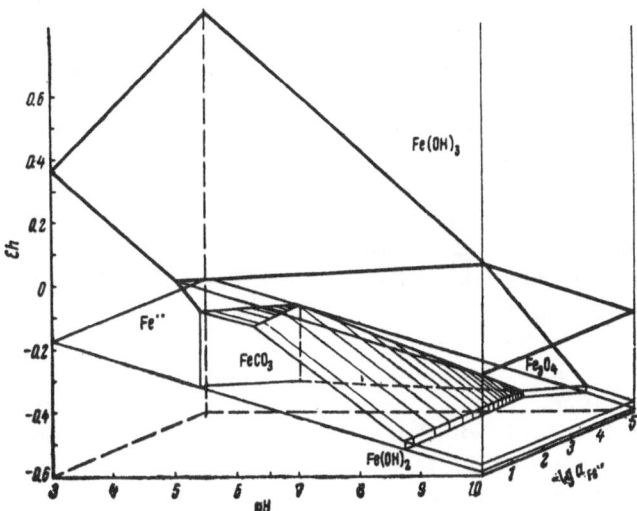

Fig. 24. Stability regions of $Fe^{\cdot\cdot}$, $Fe(OH)_3$, Fe_3O_4, $FeCO_3$, and $Fe(OH)_2$ at $\Sigma CO_2 < 8.5 \, 10^{-5}$ mole/liter.

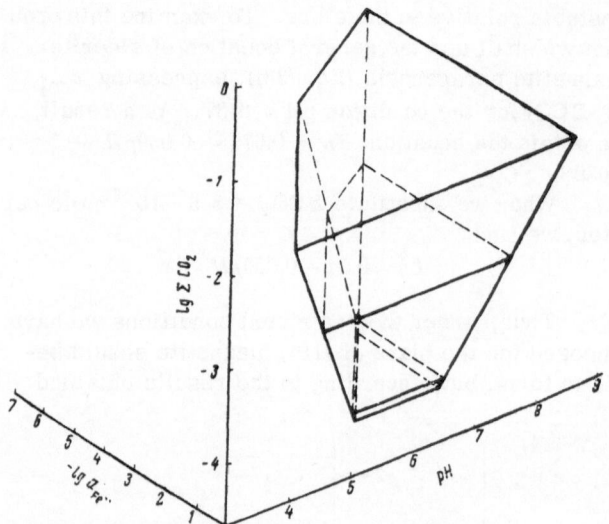

Fig. 25. Dependence of the stability zone of siderite on log ΣCO_2.

for this plane, magnetite becomes unstable relative to siderite. Such numerical results should be expected, since, in essence, we selected the condition when the $Fe(OH)_3 - Fe_3O_4$ and $Fe(OH)_3 - FeCO_3$ planes coincide. This example was chosen in order to make the fact clear that magnetite does not form when $\Sigma CO_2 = 8.5 \cdot 10^{-5}$ mole/liter. The stability diagram for iron minerals with consideration of this detail is shown in Fig. 23.[†] The case for $\Sigma CO_2 < 8.5 \cdot 10^{-5}$ mole/liter is shown in Fig. 24. This method does not give a complete picture of the changes in the stability zone of siderite as a function of ΣCO_2. To learn more of the details concerning the stability zone of siderite as it depends on changes in ΣCO_2, it is necessary to exchange coordinate values, introducing instead some variable values of ΣCO_2. In our example it would be most convenient to replace Eh by ΣCO_2. There is a very simple method for making this substitution.

Thus, in particular, taking the two equations that define the two adjoining zones of mineral stability according to Eh and setting them equal to each other makes it possible to rid ourselves of this variable and to consider the stability boundaries according to pH, $a_{Fe\cdot\cdot}$, and ΣCO_2. Let us use this method for more detailed study of the paragenetic boundaries for siderite–magnetite and siderite–Fe$^{\cdot\cdot}$. After equating Eqs. (18) and (31) according to Eh, we find that pH = log ΣCO_2 + 10.51.

The equating of Eqs. (30) and (31) with Eq. (13) according to Eh, within the plotting limits of the diagram, permits us to find, respectively,

$$pH < 6.37; \quad \lg a_{Fe\cdot\cdot} = 6.05 - \lg \sum CO_2 - 2pH$$

$$pH > 6.37; \quad \lg a_{Fe\cdot\cdot} = -0.39 - \lg \sum CO_2 - pH.$$

From these data we have constructed a three-dimensional diagram (Fig. 25), which must be interpreted in the following way: when the oxidation–reduction potential drops so low that it becomes possible for siderite or magnetite to separate out in the solid phase, then, within the limits of the parameters shown on the graph, it will be siderite that forms. Magnetite cannot form under these conditions. It will form to the left of siderite under more reducing conditions. With decline in ΣCO_2, the zone in which siderite forms under more oxidizing conditions than magnetite diminishes progressively, and, finally, when $\Sigma CO_2 < 8.5 \cdot 10^{-5}$ mole per liter, siderite may form only in an environment more reducing than that for magnetite.

In order to conclude our investigation of the given system under closed conditions, it is necessary to consider still one other paragenetic plane: $FeCO_3 - Fe(OH)_2$. The paragenesis of these compounds does not depend on the value of Eh, since, for either to form, Eh must be sufficiently high to satisfy the solubility product of one compound or the other. We therefore suggest the following series of computations:

$$\frac{a_{Fe\cdot\cdot} \cdot a_{CO_3''}}{a_{Fe\cdot\cdot} \cdot a_{(OH)'}^2} = \frac{a_{CO_3''}}{a_{(OH)'}^2} = \frac{2.2 \cdot 10^{-11}}{1.4 \cdot 10^{-15}} = 1.57 \cdot 10^4 \quad (34)$$

Expressing $a_{(OH)'}^2$ by a_H^2. and $a_{CO_3''}$ by ΣCO_2, we find for the different pH zones

$$3 \leqslant pH < 6.37, \quad \lg \sum CO_2 = -6.9$$

$$6.37 < pH < 10.33, \quad \lg \sum CO_2 = pH - 13.3$$

The paragenetic plane was plotted on a diagram in accordance with the results obtained. An examination of this diagram leads us to conclude that even at low values of ΣCO_2 the possibility of ferrous hydroxide forming in a closed system is restricted to a very narrow zone, lying in a strongly alkaline environment.

Diagrams for open systems may be plotted using Eqs. (21) and (17):

$$Eh = 0.466 - 0.059 pH + 0.059 \lg P_{CO_2}, \quad (35)$$

For an acidic environment, we find from Eqs. (21) and (16)

$$\lg a_{Fe\cdot\cdot} + \lg P_{CO_2} + 2pH = 7.47$$

For alkaline conditions, more specifically, for the conditions at the transition from magnetite

---

[†] All computations for plotting this diagram have been omitted, since they are similar in character to those described for $\Sigma CO_2 = 10^{-2}$ mole/liter.

to the stability field of siderite, using Eqs. (21) and (33), we find

$$Eh = 0.548 - 0.059 pH + 0.088 \lg P_{CO_2}$$

The siderite−ferrous hydroxide paragenesis is considered on the basis of Eq. (34) in a manner similar to that used in investigating paragenesis according to ΣCO_2. ΣCO_2 is here replaced by P_{CO_2} according to Eq. (21). After transformations it becomes clear that the occurrence of the minerals $FeCO_3$ and $Fe(OH)_3$ is possible at $P_{CO_2} = 3.39 \cdot 10^{-6}$ atm. Consequently, siderite forms at $P_{CO_2} >$

$3.39 \cdot 10^{-6}$ atm, but the formation of ferrous hydroxide is limited by the conditions $P_{CO_2} < 3.39 \cdot 10^{-6}$, atm, which actually do not exist. In Figs. 26, 27, and 28, we have shown the particular cases of mineral paragenesis for the given values of P_{CO_2}.

In comparing the diagrams plotted on the principle $\Sigma CO_2 = $ const (see Figs. 22, 23, and 24) and $P_{CO_2} = $ const (see Figs. 26, 27, and 28), one can but note the great difference in positions of the stability zones of the minerals within the ranges of pH. The essential point of the difference lies in the complex function relations among P_{CO_2}, ΣCO_2, and pH,

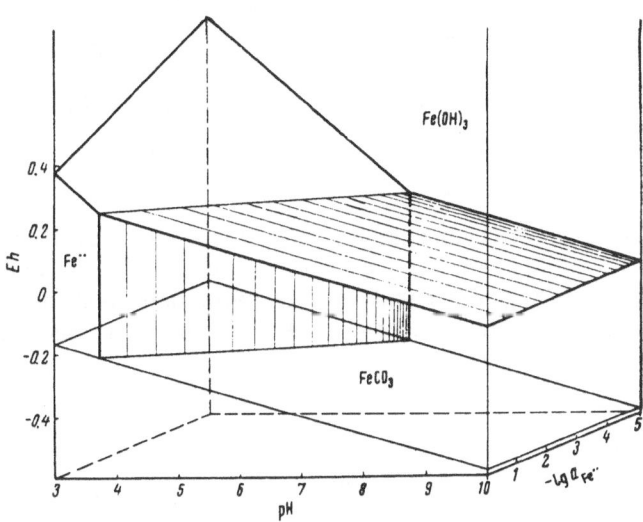

Fig. 26. Stability regions of Fe··, Fe(OH)₃, and FeCO₃ at $P_{CO_2} = 1$ atm.

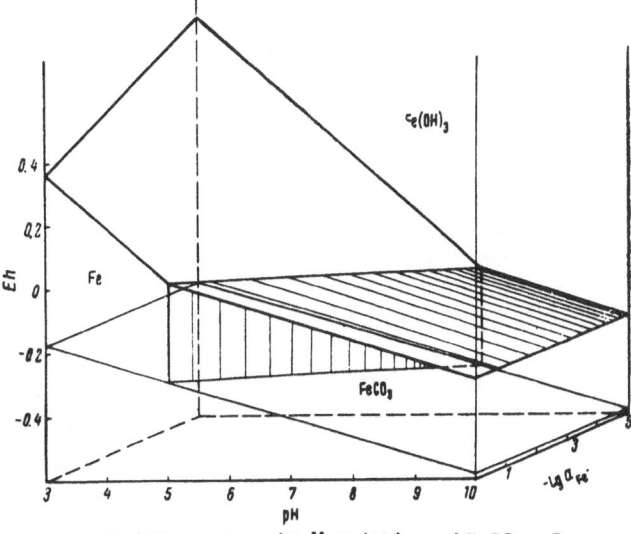

Fig. 27. Stability regions of Fe··, Fe(OH)₃, and FeCO₃ at $P_{CO_2} = 2.55 \cdot 10^{-3}$ atm.

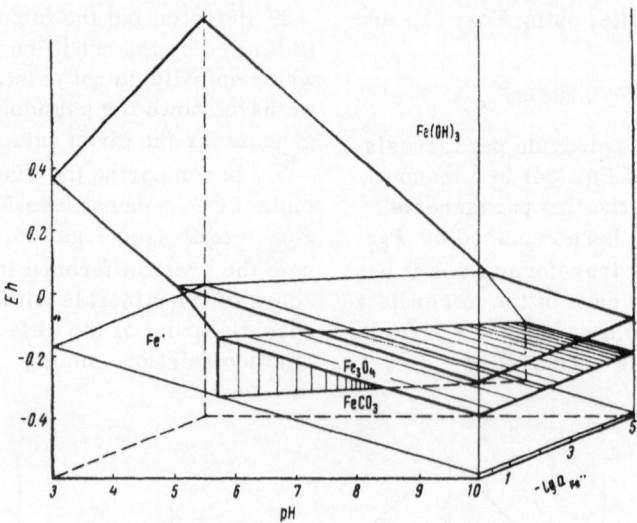

Fig. 28. Stability zones of Fe$^{··}$, Fe(OH)$_3$, FeCO$_3$, and Fe$_3$O$_4$ at
$P_{CO_2} = 10^{-4}$ atm.

which we have already discussed. Only by perceiving them in their generalized form, and not applying them to particular values of ΣCO_2, is it possible to grasp the fact that there is no difference between them. If we consider them from the viewpoint of particular cases, as was done by Garrels or Huber [Huber and Garrels, 1953], we may come to the erroneous conclusion that the zone of siderite occurrence is restricted to only acidic environments. We therefore prefer to use, in our further discussions, constructions relative to the partial pressure of any gas, and diagrams plotted on the principle of constant total activity of the gas will be resorted to only in necessary cases.

It should be noted, in general, that diagrams in pH–Eh coordinates are of little use in analyzing authigenic mineral growth in marine sediments. It is convenient to use them only for situations in which such parameters as pH and Eh vary between wide limits. The use of Eh–pH diagrams for investigation of mineral formation and paragenesis in marine sediments is a tradition rather than a necessity, since the pH value during sedimentation and diagenesis changes very insignificantly (1 to 1.5 units).

It is especially important to note that during the stage of sediment accumulation in the sea, minerals form under conditions that approach P_{CO_2} = const, since microbiological processes constantly replenish the CO_2. In the diagenetic stage the process moves more rapidly in accordance with the conditions ΣCO_2 = const. The nature of the changes

in carbon-dioxide equilibrium for investigating the formation of minerals in marine sediments is therefore more important than the amount of change of pH. For better representation of the process it is necessary to plot either ΣCO_2 or P_{CO_2} on the diagram instead of pH. In the present example this has no significance, since ΣCO_2 is related to P_{CO_2} only by pH. And, since the diagrams are plotted for pH = const, Eq. (22) is transformed to

$$\Sigma CO_2 = K \cdot P_{CO_2}$$

where K is a constant.

The construction of such diagrams by means of the equations is not difficult. The graphs illustrate two particular diagrams for the two cases of pH = 6 (Fig. 29) and pH = 8 (Fig. 30). The latter

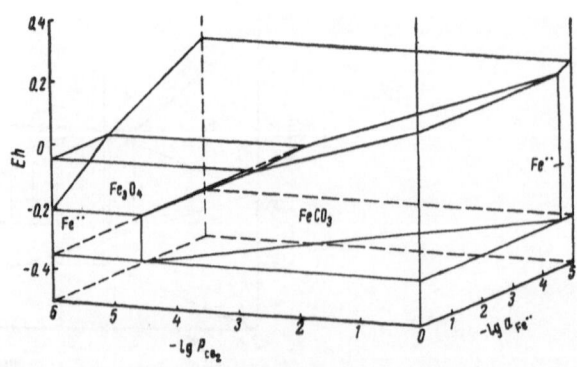

Fig. 29. Stability of iron minerals in the absence of the sulfide ion (pH = 6).

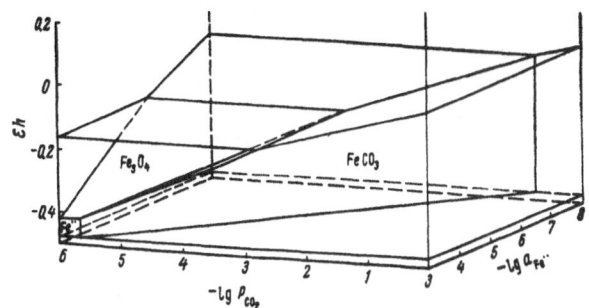

Fig. 30. Stability of iron minerals in the absence of the sulfide ion (pH = 8).

reflects marine conditions. In comparing these diagrams, we may see a difference in character. The difference involves the fact that the diagram for pH = 6 was plotted for the range of $\lg a_{Fe^{..}}$ from 0 to −5 along the z axis, whereas the diagram for pH = 8 is for the range from −3 to −8. The reason for this appears from the following considerations.

In accordance with Eq. (12), $\lg a_{Fe^{..}} = 12.94 - 2pH$ and, consequently, the greatest possible logarithm of ferrous-ion activity in a solution with pH = 6 is 0.94, but in a solution with pH = 8 the activity is −3.06 (activity expressed in moles per liter), which is natural, since hydrolysis increases with rise in alkalinity. This aspect of the behavior of the ferrous ion in solutions makes it necessary to introduce two new concepts, useful in investigating the conditions under which minerals form: a medium with high concentration of the ferrous ion and a medium with low concentration of this ion. Then, at pH = 8 the content will be high; at pH = 6 it will be low.

For geologists, however, this plays no significant role, since the contents of ions are determined from the paragenesis. This division makes it possible to describe the general features of the diagrams independently of pH. The medium with a high ferrous-ion concentration is characterized by the possible paragenesis of three minerals depending on pH and P_{CO_2} (or ΣCO_2). In the zone of high CO_2 content (or high carbon-dioxide pressure), two minerals may exist: ferric oxide and siderite. In the zone of low CO_2 pressure in the sediments, depending on Eh, three minerals may be found, replacing each other with decline in Eh: ferric hydroxide, magnetite, and siderite. In this case a definite correspondence is maintained between the sequence of minerals that form and the $Fe^{..}/Fe^{...}$ ratio known from the literature. In the zone of high CO_2 pressure, however, this ratio becomes unreal, since

$Fe(OH)_3$ immediately goes to $FeCO_3$ with declining Eh. One's attention is captured by the fact that, in principle, the possibility of siderite forming appears under conditions more strongly oxidizing than the possibility of magnetite forming. It is this sequence that is most frequently observed in nature. It disrupts the sequence of change in $Fe^{..}/Fe^{...}$ with change in Eh. In this connection, the widely publicized series of mineral indicators proposed, in particular, by Teodorovich [1947, 1956, 1964] in the form

pyrite or marcasite	siderite and iron (ferrous) chlorite	iron chlorite (ferrous−ferric), magnetite	glauconite, iron chlorite (chiefly ferric)	ferric hydroxides and oxides

cannot be recommended as a standard for determining the reducing character of the environment in which sediments are forming. Siderite and magnetite change places in this series according to ΣCO_2 (or P_{CO_2}). As will be seen from our later discussions, other members of this series may also exchange places depending on specific conditions.

A medium with a low ferrous-ion concentration is characterized by only two minerals: ferric hydroxide and siderite. Thus, magnetite in marine sedimentary rocks appears, first, as an indicator of low oxidation potential, second, as an indicator of low CO_2 partial pressure, and, third, as an indicator of an environment with high ferrous-ion concentration.

As already pointed out, the restriction of the natural environment furnishes us grounds for stating that ferrous hydroxide cannot form in nature. The use of $Fe(OH)_2$ as the initial compound for various genetic schemes may therefore lead to misunderstanding. As an example, we may cite the computations of Gulyaeva [1956]. This investigator, in using Pustovalov's [1940] model of equations, suggested the following reactions to explain the possible joint occurrence of ferous hydroxide (!) and siderite:

$$4Fe(OH)_3 + C = CO_2 + 4Fe(OH)_2 + 2H_2O$$
$$Fe(OH)_2 + CO_2 = FeCO_3 + H_2O$$

"Thus," wrote Gulyaeva [1956], "according to this reaction, only one-quarter of the reduced iron will form siderite, and three-fourths will remain in the hydrate form." With a small amount of oxygen present, the reaction proceeds according to the scheme

$$2Fe(OH)_3 + C + O = 2Fe(OH)_2 + CO_2 + H_2O$$
$$2Fe(OH)_2 + CO_2 = Fe(OH)_2 + FeCO_3 + H_2O$$

From this it is concluded that under these conditions half the iron is bound in siderite, and half remains in the ferrous-hydroxide form. With a comparatively large amount of oxygen, the following reaction may be suggested:

$$2Fe(OH)_3 + 2C + 3O = FeCO_3 + 3H_2O$$

Thus, even if we step aside from the dispute concerning the formal validity of the equations, the participation of $Fe(OH)_2$ as a solid phase gives us no right to adopt these equations as the basic scheme.

Similar remarks apply also to the equations proposed by Pavlov [1964], who, to explain the conditions under which siderite forms, stipulated $Fe(OH)_2$ in the solid phase, on very weak grounds, and on this basis made his thermodynamic computations.

For such reactions to take place in sedimentary rocks it would be necessary to propose an unbelievably strong reducing environment in the practical absence of CO_2 but with the subsequent effect of carbon dioxide that forms through hypothetical processes.

The Appearance and Behavior of Sulfide Ions in Sediments

In examining the stability fields of ferric hydroxide and magnetite, and the paragenesis of these minerals, we have used only three variables, varying independently of each other: Eh, pH, and $a_{Fe}\cdots$. The succeeding stage, investigation of the stability fields of siderite, required the introduction of one more variable: the partial pressure of CO_2, or ΣCO_2. In investigating the stability fields of siderite we may convince ourselves of the fact that there is no basis for assigning to siderite value as an indicator depending only on Eh and pH, even as there is not always a basis for concluding that the Eh during formation of a mineral may be defined by the ferrous—ferric ratio in a given mineral complex.

Let us consider a system by means of which it will be possible to investigate the stability field of iron sulfides. At first glance it may be shown that this is a simpler system than the system siderite—ferric hydroxide, since both the ferrous ion and the sulfide ion, as components of iron sulfide, are dependent on Eh and pH. They are also related by the solubility product. Since there is a definite connection between the two ions, and the ions themselves are determined by the same variables, the entire system may be reduced to the relationship between two variables: Eh and pH. Iron sulfide then naturally becomes a very important indicator of Eh and pH in the environment. Until now it has been treated in just this way by many foreign investigators (R. M. Garrels, N. K. Huber, K. B. Krauskopf, and others). Their ideas have been carried over into the domestic literature on lithology. The lack of agreement between calculated results and actual data from recent sediments by no means disturbs these investigators, who consider sulfides to be indicators and who ordinarily explain the disagreement by the complexity of natural systems and by a whole series of unrecognized facts.

Detailed analysis has shown, however, that the system of iron sulfide, under conditions of equilibrium, is more complex than the system just examined, and the role of sulfides as indicators of Eh and pH is therefore very doubtful, having even less basis than the role of siderite. In order to analyze this problem we must first consider the behavior of sulfur in aqueous basins.

The question of a biological origin of the sulfide ion in interstitial water of muddy bottoms was raised as early as the end of the last century. Andrusov [1892] and Lebedintsev [1892] proposed the possibility of hydrogen sulfide as a result of reduction of sulfates, emphasizing, it is true, purely chemical reduction by means of organic substances. Zelenskii [Zelenskii and Brusilovskii, 1893] confirmed this experimentally, but he ascribed the reduction to microbiological processes with the aid of the bacillus *Bacterium hidrosulfureum Ponticum*. Somewhat later this idea was confirmed by Egunov [1901]. The existence of such bacteria has been disputed, however, since the indicated species may grow under both aerobic and anaerobic conditions. It is likely that this form actually included several different species of bacteria [Tauson, 1932]. The question of the microbiological origin of hydrogen sulfide in muds of aqueous basins ceased to raise doubts after the experiments of Beijerinck [1895], who discovered channels in the mud and separated a pure culture of sulfate-reducing bacteria *Spirillum desulfuricans* (subsequently named *Microspira desulfuricans*).

Since that time the list of species of sulfate-reducing bacteria has been greatly extended, and the problem of energy sources for their existence has been worked out rather well, but the following remains unclear: How is the thermodynamics of hydrogen sulfide arising by microbiological means related to the environment in which the hydrogen sulfide has formed? Whereas this question has no fundamental significance for microbiologists, for geologists it is basic, since only by solving it will we properly understand the conditions under which sulfides form in sedimentary rocks.

At the present time a large number of geologists are satisfied with the view that the origin of sulfide ions is associated with anaerobic microorganisms and, consequently, that it is effective under reducing conditions in the absence of free oxygen. This rather inaccurate view cannot be reliably evaluated by us if we maintain the relative concept of "oxidizing" and "reducing" conditions and the corresponding views of aerobic and anaerobic microorganisms. The single term anaerobic, which means development under reducing conditions, still does not characterize the environment. This is due to the fact that the concept "anaerobic" itself is somewhat vague. Beginning with facultative anaerobic organisms and ending with obligate anaerobic organisms, an environment may range from 0 to 20 rH$_2$, and even the concept of "strict anaerobic organism" spreads over the range 0–12 rH$_2$ [Kanel', 1937] or 0–18 rH$_2$ [Rabotnova et al., 1955]. In particular, exclusively anaerobic sulfate-reducing microorganisms even at rH$_2$ = 17 produce hydrogen sulfide with a concentration of 0.012 mg/liter [Ivanov, 1960]. Weak development of hydrogen sulfide (1 on the intensity scale) is observed even at rH$_2$ = 24 [Kuznetsova, 1961].

It is possible that still another circumstance is effective here, that particular microorganisms may create environments differing in reducing capacity according to the potential-limiting system. Therefore, since we have resolved to construct the work on the actual figures in order to shed light on the special aspects of hydrogen-sulfide formation and the behavior of this substance over a wide range of conditions, it is necessary for us, at least at first, to avoid the amorphous concepts of "oxidizing" and "reducing" environments and the corresponding concepts of "aerobic" and "anaerobic" microorganisms.

For actual comparisons it is better to use numerical data. For this it is necessary to select some level of oxidation–reduction potential in relation to which we may consider the life conditions of sulfate-reducing bacteria. In 1932 Tauson showed that, formally, the reaction between sulfates and methane may take place under normal conditions with the formation of hydrogen sulfide. This formal possibility has served as the basis for the activity of sulfate-reducing bacteria.

Pel'sh [1937], having studied the energy relations of the sulfate-reducing process, among several schemes to explain the energy possibilities of heterotrophic microorganisms, outlined the following scheme:

$$C_6H_{12}O_6 + 3H_2SO_4 \longrightarrow 6CO_2 + 3H_2S + 6H_2O$$
$$\Delta G = -180.98 \text{ kcal}$$

If we examine the elemental act of this reaction relative to sulfur, we are led to the equation

$$SO_4'' + 10H^{\cdot} + 8e = H_2S + 4H_2O$$

For autotrophic supply [Sorokin, 1954], the process in general is presented in the form

$$SO_4'' + 4H_2 \longrightarrow S'' + 4H_2O$$

In the microbiological literature appear a large number of schemes to explain the mechanics of the sulfate-reducing process. In these, different intermediate stages of sulfur reduction are distinguished: the stages of sulfite formation [Postgate, 1951], thiosulfate formation, and tetrathionate formation [Young and Maw, 1958]. Ultimately, however, the life process of sulfate-reducing microorganisms reduces to the elemental reaction

$$SO_4'' + 4H^{\cdot} + 8e = S'' + 4H_2O$$

Therefore, there is no necessity for us to go deeper into the theory of this process, to examine the principal and minor discrepancies in the various schemes discussed by microbiologists. For us it is important simply that the theory combines the schemes and, consequently, causes no dispute. Apart from the indicated reaction, attesting to the fact that the reduction of the sulfate ion always proceeds to conclusion, i.e., to the most highly reduced form of sulfur, still another, and perhaps more important, detail is characteristic: sulfate reduction with all sulfate-reducing microorganisms does not go beyond the stage of elemental sulfur. From the formal point of view, therefore, for supplying a basis for the formation of the sulfide ion, we may agree with the representation of the reaction in the form

$$SO_4'' + 10H^{\cdot} + 8e = H_2S + 4H_2O \qquad (36)$$

The ordinary formal way of computing yields the following equation of the oxidation–reduction potential:

$$Eh = 0.303 - 0.075pH + 0.0075 \lg \frac{a_{SO_4''}}{a_{H_2S}}$$

According to the data of Sverdrup and Fleming, used by Krumbein and Garrels [1952] in their calculations, for recent marine sediments

$$H_2S + SO_4'' \approx 3 \cdot 10^{-2} \text{ moles/liter}$$

The equations of the oxidation—reduction may then be written in the form

$$Eh = 0.303 - 0.075 pH + 0.0075 \lg\left(\frac{3 \cdot 10^{-2}}{a_{H_2S}} - 1\right)$$

and, assigning different values of a_{H_2S}, we obtain Table 5.

One should expect that the data in Table 5 will describe the level of optimal activity of sulfate-reducing microorganisms under present-day marine conditions, since it excludes the necessity of expending energy on the reduction of sulfates. The plane obtained from these data on the graph is therefore called the theoretical plane of optimal activity of the microorganisms. It is shown that the actual plane of optimal conditions for the existence of sulfate-reducing bacteria is in a more highly oxidizing zone than the plane obtained by theoretical calculation.

Thus, according to Baas Becking and Kaplan, the most favorable conditions for development of sulfate-reducing bacteria are those with a pH of 6.2 to 7.9, and an Eh of −50 to −150 mV. Somewhat

lower limits are given by Kuznetsov, et al. [1962]. According to Kuznetsov the optimal conditions for development of the anaerobic sulfate-reducing bacteria *Vibrio desulfuricans* are characterized by $rH_2 = 8$, but in all the typical calcium–sulfate waters of the Carpathian region, where several types of microorganisms are active, intense production of hydrogen sulfide is observed at $rH_2 = 12$.

We might cite a whole series of facts indicating inability to compare the microbiological process with the theoretical plane of optimal activity of sulfate-reducing bacteria. For example, the data from Lake Belovod', taken from diagrams published in the work of Kuznetsova et al. [1962], indicate that at $rH_2 = 16$ the content of hydrogen sulfide amounts

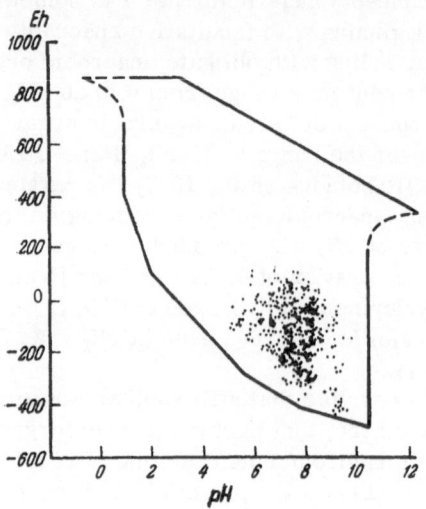

Fig. 31. Eh vs pH, parameters of conditions for existence of sulfate-reducing bacteria (after Bass Becking, Kaplan, and Moore [1960]).

TABLE 5. Relations of pH, Eh, and a_{H_2S} for Eq. (36)

a_{H_2S}, moles/liter	Particular equation
10^{-1}	$Eh = 0.302 - 0.075 \; pH$
10^{-5}	$Eh = 0.329 - 0.075 \; pH$
10^{-10}	$Eh = 0.369 - 0.075 \; pH$
10^{-15}	$Eh = 0.404 - 0.075 \; pH$
10^{-20}	$Eh = 0.417 - 0.075 \; pH$
10^{-25}	$Eh = 0.479 - 0.075 \; pH$
10^{-30}	$Eh = 0.516 - 0.075 \; pH$

TABLE 6. Content of Hydrogen Sulfide and Intensity of Its Formation in Waters of the Rozdol Sulfur Deposit. Summer of 1958 (after M. V. Ivanov [1960])

Borehole No.	pH	Eh, mV	$H_2S \cdot 10^3$ moles/liter	$SO_4'' \cdot 10^3$ gram-ion/liter	Intensity of H_2S formation mg/liter
2 c	7.0	+0.026	0.50	9.74	0.012
9 a	7.1	−0.097	1.56	10.78	1.254
8	7.1	−0.094	1.60	11.56	0.173
8 c	7.4	+0.201	0	0.56	0
9 c	7.5	+0.092	0	—	0
16	7.0	+0.028	0.90	10.17	0.294
5	6.9	−0.004	0.98	13.50	2.015
19 a	6.8	−0.033	1.13	13.87	0.126
30	7.1	−0.091	1.53	9.73	0.029
150 c	7.3	+0.019	0.73	1.84	0.009
24	7.2	−0.010	0.67	14.34	0.692
31 b	7.2	−0.107	1.60	13.66	1.485
32	7.0	−0.108	1.92	13.64	1.435

to about 10^{-15} mole/liter, which appreciably exceeds the normal content corresponding to the oxidation—reduction potential calculated from Eq. (36).

In Fig. 31 we have shown the region in which reduction of sulfate takes place, or, more properly, where the sulfide that has formed is again stable [Baas Becking et al., 1960]. Judging from Fig. 31, the region where sulfate-reducing bacteria are active extends far beyond the limits of hydrogen-sulfide stability, in accord with the thermodynamics of the elemental reaction of sulfate reduction, Eq. (36).

In Table 6 we have presented data from Ivanov [1960], recalculated by us, on the intensity of the sulfate-reducing process in waters of the Rozdol sulfur deposit. As seen in column three of this table, the values of Eh range rather widely. The corresponding values of H_2S, given in the next column, indicate that the hydrogen-sulfide content, even when the activity coefficient is considered, greatly exceeds the contents theoretically calculated from (36) and do not entirely correlate with the Eh values. In comparing these data with the data of the next column, we note that, even at high values of the oxidation—reduction potential in a known oxidizing environment, the possibility of microbiological reduction of sulfates is high.

Active reduction of sulfates is observed in the Black Sea at a depth of 300 m, i.e., in the upper part of the hydrogen-sulfide zone, where the oxidation—reduction potential is rather high [Skopintsev, 1957]. Such reduction was proposed by Skopintsev [1953] and was confirmed by Sorokin [1962]. It is clear that the opinion that anaerobic microorganisms begin to manifest activity only under reducing conditions is somewhat inaccurate. Proof of this is found in the experiments of Rabotnova [Rabotnova et al., 1955]. She placed a culture of pure putrefactive anaerobic bacteria in a medium with $rH_2 = 22$. In the course of several hours these bacteria lowered the medium to $rH_2 = 3-5$.

Considerable experience in the study of the interrelations between the microorganisms and their environment led Rabotnova [1957] to the conclusion that anaerobic organisms have a very clearly defined capacity for adapting the environment to their requirements, that they may develop under aerobic conditions and, because of their activity, lower the oxidation—reduction potential. In regard to sulfate-reducing microorganisms, similar relations were brought to light by V. I. Aleshina as early as 1938. She showed that, on pure cultures of sulfate-reducing bacteria, sulfate reduction begins in a neutral medium at positive values of Eh (+169, +184, and others).

The same conclusions may be arrived at from the experiments of Rozhkova [Rozhkova et al., 1965]. "Optimal conditions" must therefore be considered not as corresponding to an environment with a particular Eh in which sulfate-reducing bacteria are developed most intensely but as corresponding to an environment created because of the intense development of sulfate-reducing bacteria. This means that, because of the intense activity of the indicated microorganisms, hydrogen sulfide is actively formed, and this substance, thanks to its readily exchanged electrons, facilitates the lowering of the oxidation—reduction potential. Ivanov [1961] came closest to an understanding of this process. The proposed formulation in great measure vindicates the view that the sulfate-reducing process may take place rather effectively at high values of Eh.

Common sense thus prompts us to state that numerals characterizing the zone of optimal development of sulfate-reducing anaerobic bacteria must be somewhat greater than those normally used by microbiologists. In order not to depart from the facts, however, we have assigned rigid conditions, as set down by the experimenters, and we have used for plotting the actual planes of optimal activity of sulfate reducing microorganisms the values $rH_2 = 10-12$. As seen from Fig. 32, the zone corresponding to these values lies above the theoretical plane of optimal activity. But these planes also prove to have no connection with the amount of hydrogen-sulfide production. Thus, according to L. D. Shturm, the formation of hydrogen sulfide reached 700 mg/liter at optimal development of sulfate-reducing bacteria, and in saline basins the content reached 900 mg/liter, corresponding to $H_2S = 2 \cdot 10^{-2}$ to $2.6 \cdot 10^{-2}$ mole/liter.

There is also great significance in the fact that there are aerobic (*Vibrio hydrosulfurens*) and facultative-anaerobic (*Vibrio halobius desulfuricans*, *Bacterium desulfuricans*) bacteria that may produce sulfide with the admission of air [Rubenchik, 1947], although with considerably less productivity than bacteria of the genus *Sporovibrio*.

Considering all we have discussed above, we can reach but a single conclusion: by means of sulfate-reducing bacteria, hydrogen sulfide is formed energetically by using the energy possibilities of organic material. This means that the formation of hydrogen sulfide is not controlled by the oxidation—reduction relations of the medium in which it appears. We reached this conclusion in 1963 [Stashchuk, 1963]. The same conclusion was reached the following year by Berner [1946b], who used pS" as a variable independent of Eh in plotting his diagrams. The reaction of direct reduction of the

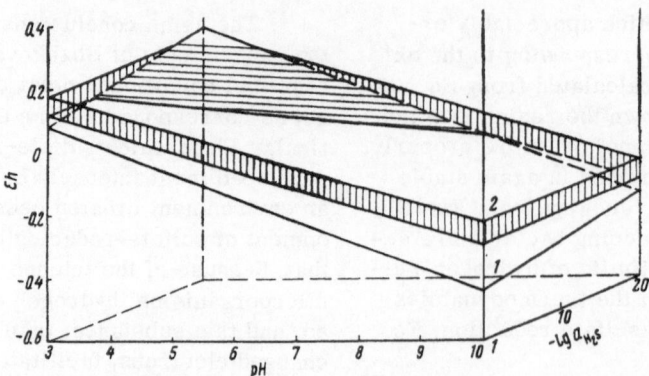

Fig. 32. Theoretical plane (1) and actual zone (2) of optimal activity
of sulfate-reducing microorganisms.

sulfate ion to hydrogen sulfide without the participation of microorganisms is not very probable in
general. Proof of this may be found in an experiment taken from the paper of Huber and Garrels
[1953, p. 337]:

"Attempts to precipitate ferrous sulfide.
Ferrous sulfate in approximately the concentration of sea water
was added to several of the solutions to see if pyrite or other
iron sulfide would be formed by reduction of sulfate under appropriate conditions. Even though the oxidation potential and
the pH dropped well within the field where sulfate reduction would
be expected (Run No. 2 at 62 h), no iron sulfate was formed. This
is fully to be expected; no examples of sulfate reduction in natural
inorganic water solutions could be found in the literature."

The fact that it is difficult to convert sulfates
to sulfides, even with appropriate lowering of the
potential, was pointed out by Baas Becking and
Moore [1961]. Furthermore, organic matter alone
cannot reduce sulfates. "... All attempts to attain
reduction of sulfates at moderate temperatures during protracted experiments in the presence of dead
organic matter, including various bitumens, proved
unsuccessful" [Ginzburg-Karagicheva, 1954, p. 151].

The suggestion that "microorganisms reduce
sulfates to H_2S only under reducing conditions [Lu-
kashev, 1963, p. 236] is the automatic consequence
of the view that the presence of hydrogen sulfide in-
disputably indicates the complete absence of oxygen,
a view advanced by V. O. Tauson as early as 1932.
The apparently logical completeness of this conclusion arouses no doubt at the present time because,
in addition to the other data, oxygen has not been
detected analytically in the presence of hydrogen
sulfide in the deep part of the Black Sea or on the
continental shelfs.

However, since in this section we are considering basically the problem of the formation of
iron sulfides, and not the application to particular
cases, it becomes important to discover if the con-

clusions have a universal character. It has been
shown that the existence of hydrogen sulfide does
not always indicate complete absence of oxygen and
that oxygen may exist in small quantities in a reducing hydrogen-sulfide environment, but giving no
essential indication of the oxidation–reduction potential of the environment.

During investigations of the Caspian Sea
[Klenova, 1956], it was noted that at shallow depths
the bottom layers of the Krasnovodsk Gulf, which
contain fairly large amounts of oxygen, also contain
amounts of hydrogen sulfide (0.93–0.86 cm^3/liter)
that are high for the Caspian, whereas in other parts
of the gulf, at greater depths, where oxygen is practically absent, the content of hydrogen is appreciably
lower. In the region where the oxygen and hydrogen-
sulfide zones of the Black Sea come together, there
occurs a zone up to 35 m thick where oxygen and
hydrogen sulfide exist together [Gololobov, 1953].
The process of hydrogen sulfide formation in the
presence of oxygen also characterizes muds. Using
the intensity of the microbiological sulfate-reducing
process in the muds of Lake Belovod', with tracer
atoms of S^{35}, Ivanov [1956] obtained the following
results:

Mud from a depth of 25 m, the hydrogen-sulfide zone. The intensity of H_2S formation amounts
to 0.128 mg/liter per day without use of sucrose
and 0.114 mg/liter per day with sucrose as a food
supplement.

Mud from a depth of 9.5 m, the oxygen zone.
Reduction amounts to 0.067 mg/liter per day without sucrose and 0.097 mg/liter per day with use of
sucrose.

These data indicate that the presence of hydrogen sulfide still does not necessarily mean that
oxygen is absent in the gaseous phase in muds and
bottom waters. Complete absence of oxygen is not

necessary for reduction of sulfate; it is necessary merely that the advent of oxygen be sufficiently small [Ginzburg-Karagicheva, 1954]. Below we shall define more accurately the conditions under which hydrogen sulfide exists together with oxygen and under which oxygen disappears, since these determine the character of mineral formation. In the present case, however, we can draw the following conclusion, without contradiction, on the basis of the discussed facts.

Hydrogen sulfide may form in sediments within a definite but rather broad range of oxidation—reduction potential. The hydrogen sulfide that forms by microbiological means may appear in an environment that is oxygen-free or in one with perceptible but naturally small contents of oxygen, depending on the nature of the mud, depth of the hydrodynamic regime, intensity of sulfate reduction, and still other factors. The hydrogen sulfide that forms is therefore not in thermodynamic equilibrium with the environment, and it begins to oxidize, striving to reach an equilibrium state. The means by which the hydrogen sulfide is oxidized determines the stable equilibrium content of sulfide ion in the environment, and this essentially determines the possibility of sulfides forming. The next question necessary to consider is therefore that concerning the nature of the oxidation of hydrogen sulfide according to its content in the medium.

First we must make clear the role in this process that belongs to sulfur bacteria. Three large physiological groups of bacteria take part in the oxidation of hydrogen sulfide: colorless sulfur bacteria, colored sulfur bacteria, and thiobacteria. In contrast to sulfate-reducing microorganisms, the sulfur bacteria are chiefly aerobic. They are active in the presence of dissolved oxygen. This category includes, in particular, the colorless sulfur bacteria, which deposit sulfur by oxidizing hydrogen sulfide by oxygen [Ivanov, 1964]. The group of colored sulfur bacteria develop under anaerobic conditions, but all representatives of this group are photosynthesizing. They may therefore develop only where sunshine penetrates. Their range of activity is restricted to layers of water and the surface of bottom mud in those basins where the floor is well lighted.

The great bulk of thiobacteria are aerobic, developing only in the presence of oxygen [Ivanov, 1964] under rather strong oxidizing conditions. In particular, *Th. thioparus,* with a sufficient amount of the sulfide ion, is most intensely developed within the range $rH_2 = 12-16$ [Kuznetsov and

Sokolova, 1960]. The lower boundary of these bacteria has to be under conditions $rH_2 = 11.6-17.6$ [Kuznetsova, 1961].

These conditions necessarily presuppose the presence of both hydrogen sulfide and oxygen. It is no accident, therefore, that Sokolova [1961], having studied a great amount of material on the distribution of thiobacteria in stratal waters, concluded that these microorganisms are found only where deep waters come in contact with surface waters, i.e., where there is contact with oxygen. Thiobacteria have therefore been distinctive indicators: "the presence of thiobacteria in acid-free waters is apparently an indication of water-exchange capacity and openness of the structure of the given deposit" [Kuznetsov et al., 1962].

The greatest activity of thiobacteria in the Black Sea occurs at depths of 180—200 m [Sorokin, 1954], where the oxidation—reduction potential is characterized by values generally more than 100 mV [Skopintsev, 1957]. The discovery of thiobacteria in the hydrogen-sulfide zone of the Black Sea at depths greater than 2000 m may indicate something peculiar in this connection.

One's attention is drawn to the specific character of the distribution of these bacteria: they are concentrated in a very thin surficial film of muds [Kriss, 1959], just as in the oxygen zone they penetrate into the muds [Isachenko and Egorova, 1939]. If we take into account the effect of the Bosporus, that heavy oceanic waters pass through the strait into the Black Sea and move across the floor [Bogdanova, 1959, 1960, 1961; Serpoyan, 1966], the cause of the thin bacterial film of thiomicroorganisms at the surface of the mud then appears to have a simple explanation: a small amount of oxygen, necessary for the activation of thiobacteria, moves in with oceanic waters over the floor of the Black Sea.

The circulation of water in the hydrogen-sulfide zone, discovered by Skopintsev and Smirnov [1965] also apparently has some effect on this process. Rodina [1965] noted that bacteria that oxidize hydrogen sulfide occur at the boundary between the hydrogen-sulfide and oxygen layers. Lees [1955] concluded that, in those parts of a basin in which anaerobic conditions prevailed and where sulfates were being reduced, parallel oxidation of the sulfides that formed in the process may take place only by photosynthesizing bacteria.

These examples direct one's mind to the fact that, except for photosynthesizing bacteria, which apparently cannot affect diagenetic processes, all other forms of sulfur bacteria manifest activity only under conditions when hydrogen sulfide may be oxidized independently. Direct proof of this is found in the experiments of Ivanov [1957, 1959]. To explain the degree of oxidation of hydrogen sulfide in a bacterial environment, investigations were conducted with the tracer atom S^{35}, leading the author to conclude that in Sernoe Lake thiobacteria oxidize 30% of the sulfides each day, colored sulfur bacteria 8%, and, along with this, oxidation by abiogenic means takes place (17%).

As investigations by microbiologists have shown, the oxidation of hydrogen sulfide takes place in a manner fundamentally different from the reduction of sulfates. Hydrogen sulfide is oxidized by microorganisms to sulfur, and only when hydrogen sulfide is deficient is sulfur oxidized to sulfates, just as the sulfate-reducing process does not extend through the stage of elemental sulfur. Let us compare these data with data on the oxidation of hydrogen sulfide by spontaneous means.

There is an extensive literature on the oxidation of hydrogen sulfide by spontaneous means, but, unfortunately, it deals chiefly with the nature of oxidation in the presence of great oxygen excess. For these conditions, M. A. Petrov and A. K. Yakovskaya [Durov and Turzheva, 1947] showed that oxidation goes only to sulfur in an alkaline environment. Durov [1935], concerned with the purification of water of hydrogen sulfide, discovered that hydrogen-sulfide sewage water dumping into the Matsesta River immediately discards sulfur. This is due to oxidation of hydrogen sulfide according to the scheme

$$2H_2S + O_2 = 2H_2O + 2S$$

On the basis of hydrogen sulfide oxidation by chlorine, Durov and Turzheva [1947] concluded that the final oxidation product of the sulfide ion by weak oxidizing agents as well as oxygen is sulfur.

Levchenko and Makarova [1950] explained that the oxidation of sulfur to sulfates takes place after complete disappearance of sulfides. In this connection, Levchenko [1950] considered the equilibrium oxidation−reduction potential for the sulfide ion to be accounted for by the equation

$$Eh = E^0 + \frac{RT}{2F} \ln \frac{a_{S^0}}{a_{S''}},$$

or for t = 25°C

$$Eh = E^0 - 0.029 \lg a_{S}.$$

Berner [1963] came to the same conclusion from experiments on determination of the activity of the sulfide ion by means of a silver−silver-sulfide electrode.

Interesting experiments helping us to explain the nature of hydrogen-sulfide oxidation were conducted by Skopintsev [Skopintsev et al., 1961]. In oxidizing hydrogen-sulfide water with different proportions of oxygen as the oxidizing agent, this investigator became convinced that hydrogen sulfide is oxidized to sulfur in the first stage. The following reaction helps us to understand this. The oxidation of hydrogen-sulfide to sulfur may be illustrated schematically in the following manner:

$$2H_2S + O_2 = 2S + 2H_2O$$

Here, one molecule of water is needed for two molecules of hydrogen sulfide. The molecular weights of hydrogen sulfide and oxygen are almost identical, and we may say, therefore, that during oxidation of hydrogen to sulfur the weight ratio of O_2 to H_2S must be 0.5. If oxidation goes to completion to sulfates, then $O_2 : H_2S$ must be 2. This is clear from the reaction

$$H_2S + 2O_2 = H_2SO_4$$

As appears from the above experiments when narrow ratios of $O_2 : H_2S$ are involved in the reaction, the expenditure of oxygen on hydrogen sulfide is 1.0. When the initial ratio of $O_2 : H_2S$ is larger, the expenditure is 1.3. Both figures lie within the interval from 0.5 to 2; i.e., in either case oxidation does not exceed the limits of the reactions

$$H_2S \rightarrow S^0 \text{ and } H_2S \rightarrow SO_4^-$$

In connection with the fact that the formation of $S_2O_3'' + SO_3''$ was very slight and was not observed in all the experiments and that the kinetics of the oxidation of hydrogen sulfide to sulfur is characterized by a reaction of the first order, Skopintsev arrived at the following conclusion: "It is clear that the oxygen−sulfide coefficients found in our experiments reflect two stages of the process of oxidizing H_2S: a) oxidation of H_2S to finely dispersed molecular sulfur (this process was practically terminated within the observation time of each experiment); b) oxidation of sulfur to sulfuric acid (this process was not completed within the time of the experiments)" [Skopintsev et al., 1961].

Thus, the chemical course of oxidation involves two stages:

1. oxidation of hydrogen sulfide to sulfur;
2. oxidation of sulfur to sulfates.

In comparing this course of oxidation with the microbiological course, it may be seen that the two are entirely identical. On the other hand, the course of oxidation differs considerably from the course of sulfate reduction, during which there is no stage of

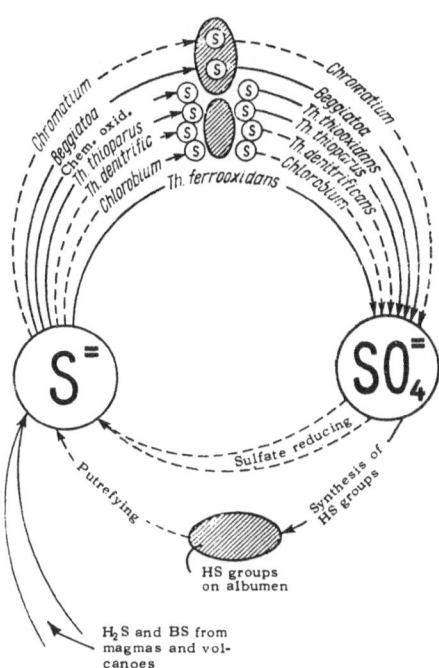

Fig. 33. The sulfur cycle with participation of
bacteria (after Ivanov [1964]).

elemental sulfur. This distinctive lag in the conversion of sulfur is well illustrated in Fig. 33, taken
from the work of Ivanov [1964].

Thus, a comparison of microbiological and
spontaneous oxidation shows that the oxidation of
hydrogen sulfide by sulfur bacteria does not disturb
the natural spontaneous process of oxidation; it
merely accelerates it. Sulfur bacteria are catalysts
of this process, leading to concentration (activity)
of the sulfide ion according to the thermodynamic
conditions of the environment. Whereas sulfate-
reducing bacteria appear as transformers, forming
hydrogen sulfide in spite of thermodynamic conditions of the environment in which it appears, sulfur
bacteria appear as dependents, adapting to the naturally occurring process, speeding up the approach
to thermodynamic equilibrium. This conclusion permits us, during thermodynamic constructions, to
ignore the biological factor of oxidation and to take
for a basis the nature of spontaneous oxidation of
hydrogen sulfide.

Unfortunately, some geologists and chemists
have not given proper attention to the obvious picture of interaction between sulfate-reducing and sulfur bacteria. To explain the formation of hydrogen
sulfide some investigators try an approach from the
position of reduction of elements according to a
series of standard potentials: at first $Mn^{...}$ is re-

duced, then $Fe^{...}$ to $Fe^{..}$, and only lastly do sulfates
begin to be reduced [Gulyaeva, 1956; Ostroumov,
1953a, 1953b; Ostroumov and Fomina, 1959; Ostroumov and Volkov, 1960; Volkov, 1961a]. This explanation completely ignores the potent factors of microbiological reduction of sulfates and supports the formal positions of Garrels concerning the possibility
of thermodynamic reduction of the sulfate ion to
the sulfide ion. On the other hand, these investigators treat the formation of elemental sulfur necessary for the formation of pyrite from the viewpoint
of the work of sulfur bacteria in a reducing environment without the admission of oxygen or light (!)
despite thermodynamics and the life capabilities of
these species of bacteria [Ostroumov, 1953a, 1953b;
Ostroumov and Fomina, 1959; Volkov, 1961; Ostroumov et al., 1961].

It would seem that the question should be
settled by the proofs we have cited. Stability diagrams of iron sulfides should be constructed on the
basis of the oxidation reaction of hydrogen sulfide
to sulfur, since "irreversible processes, because
of their lack of equilibrium, cannot be represented
on thermodynamic diagrams" [Godlevskii, 1965].
There is a circumstance, however, that requires
us still to keep our attention on this problem.

It is necessary to determine more precisely
whether the oxidation reaction of hydrogen sulfide
to sulfur is reversible. In other words, does this
reaction determine the activity of the sulfide ion in
accordance with the oxidation–reduction potential
of the environment, or, perhaps, is the activity of
the sulfide ion determined by some other intermediate product during oxidation of hydrogen sulfide to
sulfur? This question is entirely justified by the
fact that it does not always appear possible to bring
the activity of the sulfide ion in the environment in
agreement with the measured oxidation–reduction
potential.

Kryukov [1948], who believes in the existence
of the reversible reaction

$$S'' \rightleftarrows S^0 + 2e, \qquad (37)$$

after using the equation of this reaction for mineral
waters of the Caucasus, found a great divergence
between measured and computed values. He concluded that "the use of Eq. (37) in examining the
oxidation–reduction state of sulfur in mineral waters can not lead to strict quantitative conclusions
for the following reason. The standard potential
E^0 had been determined by no one with sufficient
accuracy... Sulfur may be found in different com-
binations in mineral waters. Oxidation of S'' ions

leads to the formation of the amorphous modification of sulfur S_μ, which is gradually converted to crystalline sulfur S_λ. In addition, the solubility of colloidal sulfur that forms first is greater than the solubility of coarsely dispersed sulfur. These relations limit the accuracy of computations based on Eq. (37)."

Apart from the indicated fact, we may cite the paper by Berner [1964c], who soon after assumed that the activity of the sulfide ion in a medium is determined by Eq. (37), but found it necessary after a study of recent sediments to plot stability diagrams of the iron minerals in Eh–pS″ coordinates with pH and P_{CO_2} = constant. In other words, this investigator expressed the activity of the sulfide ion as a variable independent of Eh. There exists a whole series of diagrams plotted on the basis that the activity of the sulfide ion is determined by reaction (36) and, consequently, does not depend on the oxidation–reduction potential of the sulfide-ion–sulfur system. Such diagrams first appeared in the paper by Krumbein and Garrels [1952]. The diagrams became rather widely published, including in the USSR, and continue to appear in the geological literature at the present time. Therefore, before maintaining that the activity of the sulfide ion is determined by (37), or, at least, that we have the right to use tabular thermodynamic constants, let us analyze some examples. We shall first familiarize ourselves with the behavior of hydrogen sulfide in solution.

Being dissolved in water in concentrations characteristic of natural waters, hydrogen sulfide is present chiefly in one of three forms: undissociated H_2S, HS′ ion, and S″ ion.

Thus, if the total activity of hydrogen sulfide is expressed by Σa_S, then

$$\Sigma a_S = a_{H_2S} + a_{HS'} + a_{S''}$$

The dissociation of hydrogen sulfide takes place in two stages:

1) $H^{\cdot} + HS' = H_2S, \quad \dfrac{a_{H^{\cdot}} \cdot a_{HS'}}{a_{H_2S}} = 10^{-7.001}$

2) $H^{\cdot} + S'' = HS', \quad \dfrac{a_{H^{\cdot}} \cdot a_{S''}}{a_{HS'}} = 10^{-13.891}$

Multiplying the first apparent dissociation constant by the second, we get

$$\frac{a^2_{H^{\cdot}} \cdot a_{S''}}{a_{H_2S}} = 10^{-20.892}$$

Using the resulting equation, it is possible to determine the activity of undissociated hydrogen sulfide and of each of the ions at any given value of pH:

$$a_{H_2S} = \frac{\sum a_S}{1 + 10^{pH-7.001} + 10^{2pH-20.892}}$$

$$a_{HS'} = \frac{\sum a_S}{1 + 10^{7.001-pH} + 10^{pH-13.891}}$$

$$a_{S'} = \frac{\sum a_S}{1 + 10^{20.892-2pH} + 10^{13.891-pH}}$$

On the basis of these equations, a diagram of the dominant activity of one ion or the other is plotted against the pH (Fig. 34).

As seen from Fig. 34, it may be stated with no great error that when

$$pH < 7, \quad \sum a_S \approx a_{H_2S}$$
$$7 < pH < 13.89, \quad \sum a_S \approx a_{HS'}$$
$$pH > 13.89, \quad \sum a_S \approx a_{S''}$$

Each component forms its oxidation–reduction system, which, as seen in the oxidation to sulfur, may be represented by the reactions

$$\left.\begin{array}{l} H_2S = S^0 + 2H^{\cdot} + 2e \\ HS' = S^0 + H^{\cdot} + 2e \\ S'' = S^0 + 2e \end{array}\right\} \quad (38)$$

and the corresponding equations

$$\left.\begin{array}{l} Eh = 0.142 - 0.059pH - 0.0295 \lg a_{H_2S} \\ Eh = 0.065 - 0.0295pH - 0.0295 \lg a_{HS'} \\ Eh = -0.476 - 0.0295 \lg a_{S'} \end{array}\right\} \quad (39)$$

The expression "its oxidation–reduction system" is conditional, since it is perfectly natural that, using a particular value of free energy for deriving each of the indicated equations and for determination of the apparent dissociation constant,

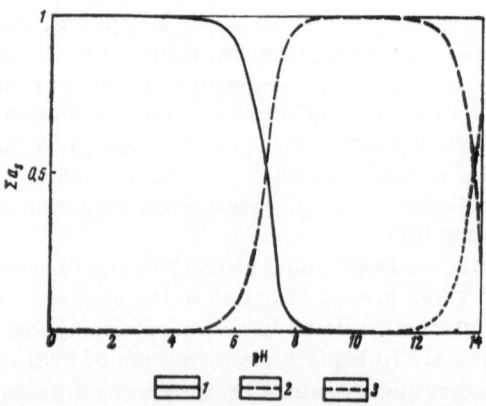

Fig. 34. Form of total activity of hydrogen sulfide according to pH. 1) a_{H_2S}; 2) $a_{HS'}$; 3) $a_{S''}$.

the choice of proper reaction acquires a purely provisory meaning and depends on the pH of the solution, for which the use of an equation is convenient. This view may be readily demonstrated in the following way. For the reaction

$$H_2S = HS' + H^{\cdot}$$

$$\Delta G^0_{reaction} = \Delta G^0_{HS'} - \Delta G^0_{H_2S}$$

$$\lg K = -\frac{\Delta G^0_{HS'} - \Delta G^0_{H_2S}}{1.364}$$

Hence

$$\frac{a_{H^{\cdot}} \cdot a_{HS'}}{a_{H_2S}} = \text{antilog}\left\{\frac{\Delta G^0_{H_2S} - \Delta G^0_{HS'}}{1.364}\right\},$$

$$\lg a_{H_2S} = \lg a_{HS'} - pH - \frac{\Delta G^0_{H_2S} - \Delta G^0_{HS'}}{1.364}$$ (40)

$$0.0295 \lg a_{H_2S} = 0.0295 \lg a_{HS'} - 0.0295 pH - \frac{\Delta G^0_{H_2S} - \Delta G^0_{HS'}}{2 \cdot 23.06}$$

In similar fashion we may calculate that for the reaction

$$HS' = H^{\cdot} + S''$$

$$0.0295 \lg a_{HS'} = 0.0295 \lg a_{S''} - 0.0295 pH - \frac{\Delta G^0_{HS'} - \Delta G^0_{S''}}{2 \cdot 23.06}$$ (41)

Let us now examine the equations of oxidation−reduction reactions

$$Eh_1 = E^0_{H_2S/S^{\bullet}} - 0.059\, pH - 0.0295 \lg a_{H_2S}$$

$$Eh_2 = E^0_{HS'/S^{\bullet}} - 0.0295 pH - 0.0295 \lg a_{HS'}$$

$$Eh_3 = E^0_{S''/S^{\bullet}} - 0.0295 \lg a_{S''}$$

E^0 is defined as $\Delta G^0_{reaction}/23.06 \cdot 2$. Therefore

$$\left.\begin{aligned}
E^0_{H_2S/S^{\bullet}} &= \frac{\Delta G^0_{S^{\bullet}} - \Delta G^0_{H_2S}}{23.06 \cdot 2} \\
E^0_{HS'/S^{\bullet}} &= \frac{\Delta G^0_{S^{\bullet}} - \Delta G^0_{HS'}}{23.06 \cdot 2} \\
E^0_{S''/S^{\bullet}} &= \frac{\Delta G^0_{S^{\bullet}} - \Delta G^0_{S''}}{23.06 \cdot 2}
\end{aligned}\right\}.$$ (42)

If we substitute a_{H_2S} in the first oxidation−reduction system in accordance with (40), it is then shown that

$$Eh_1 = E^0_{H_2S/S^{\bullet}} + \frac{\Delta G^0_{H_2S} - \Delta G^0_{HS'}}{23.06 \cdot 2} - 0.0295\, pH - 0.0295 \lg a_{HS'}$$

but, in view of system (42),

$$E^0_{H_2S/S^{\bullet}} + \frac{\Delta G^0_{H_2S} - \Delta G^0_{HS'}}{2 \cdot 23.06} = \frac{\Delta G^0_{S^{\bullet}} - \Delta G^0_{HS'}}{23.06 \cdot 2} = E^0_{HS'/S^{\bullet}}$$

In other words, we have the equations

$$E^0_{H_2S/S^{\bullet}} + 0.0295 \lg K_1 = E^0_{HS'/S^{\bullet}}$$

$$E^0_{HS'/S^{\bullet}} + 0.0295 \lg K_2 = E^0_{S''/S^{\bullet}}$$

$$E^0_{H_2S/S^{\bullet}} + 0.0295 \lg K_1 \cdot K_2 = E^0_{S''/S^{\bullet}}$$

All values of standard potentials are thus mutually derivable and do not change with change in pH.

On the basis of this, with further analysis of the oxidizing conditions of the sulfide ion, we will rest on

$$\sum a_S = a_{H_2S} + a_{HS'} + a_{S''}$$

as the value relative to which we may examine oxidation of the sulfide ion in the general form of the question and may compare this value with any standard potentials of the hydrogen-sulfide−sulfur system.

Let us now go on to an examination of experimental data on the oxidation of hydrogen sulfide. There are unfortunately few data that might have been used for precise calculations. The most interesting data have been published by Zavodnov [Kryukov et al., 1962; Zavodnov, 1965]. This author produced a table with the characteristic state of sulfide equilibrium in the waters of the Matsesta River (Table 7).

On the basis of these data, Zavodnov concluded that system (37) "determines the oxidation−reduction potential of hydrogen-sulfide mineral waters" [Zavodnov, 1965, p. 96]. This conclusion, correct in principle, leaves in force the remark of Kryukov (see pp. 59-60) and causes perplexity concerning why Berner was unable to detect this relationship in his investigation of muds. Moreover, if Eq. (37) is everywhere in force, how may we represent the activity of sulfur bacteria, which develop where the activity of hydrogen sulfide is clearly greater than necessary according to Eq. (37). It would be more nearly correct to state that "the oxidation−reduction state of hydrogen-sulfide waters even with insignificant content of hydrogen sulfide, up to 0.05 mg/liter, is better described by the system $S'' \rightleftharpoons S^0 + 2e$ than by $S'' + 4H_2O \rightleftharpoons SO_4'' + 8H^{\cdot} + 8e$" [Zavodnov, 1965, p. 97].

Whereas this conclusion is not clear from Table 7, since the computed corrections of the sulfide-ion−sulfate system are not greatly larger than the corrections for the sulfide-ion−sulfur system, rather simple analysis, which Zavodnov unfortunately did not make, leads us to reject (36) as potential determining. This analysis involves the following. For system (36)

$$Eh = E^0 + 0.0074 \lg a_{SO''} - 0.074 pH - 0.0074 \lg a_{S''}$$

for system (37)

$$Eh = E^0 - 0.0295 \lg a_{S''}$$

TABLE 7. Oxidation−Reduction Potential and the State of Sulfide Sulfur in Mineral Waters according to Zavodnov [1965]

Locality	Date of analysis	Yield, liters per day	t °C	pH	H_2S, mg per liter	SO_4, mg-mole per liter	Computed $m_{H_2S} \cdot 10^4$	$m_{HS^-} \cdot 10^4$	$a_{S^=} \cdot 10^{10}$	Eh, experimental	Eh, computed from Eq. (37)	ΔEh	Eh, computed from Eq. (36)	ΔEh
Lermontovskii No. 1 (Pyatigorsk)	30/9 1959	270	47.5	6.17	9.1	8.50	1.69	0.98	0.368	−138	−144	+6	−181	+43
Imeni Semashko (Zheleznovodsk)	10/7 1956	140	49.3	6.30	0.05	9.80	0.0082	0.0065	0.0036	−75	−78	+3	−176	+101
Talgi (Dagestan ASSR)	18/8 1958	—	35.8	6.27	381	—	72.0	39.7	12.5	−195	−203	+8		
Bh. 29 (Pyatigorsk)	6/8 1958	500	48.0	6.42	7.25	10.44	1.04	1.09	0.736	−133	−153	+20	−199	+66
The same	24/9 1959	500	43.5	6.31	7.40	10.44	1.19	0.98	0.525	−133	−148	+15	−194	+61
Bh 1 (Essentuki)	15/8 1958	250	25.1	6.57	22.1	0.01	3.67	2.81	1.16	−162	−183	+21	−207	+45
Krasnoarmeiskii No. 1 (Pyatigorsk)	7/8 1958	2.5	23.2	6.06	0.164	8.36	0.0395	0.0086	0.00102	−90	−95	+5	−131	+41
Bh 16 (Pyatigorsk)	30/9 1959	300	47	6.28	9.92	8.31	1.86	1.05	0.652	−122	−153	+31	−189	+67
» 7 »	7/8 1958	7	28.2	6.04	0.042	9.40	0.0098	0.0025	0.00034	−43	−74	+31	−130	+87
» 7 »	6/10 1959	7	28	6.08	0.042	9.40	0.0096	0.0027	0.00039	−40	−75	+35	−133	+93
» 11 »	20/8 1959	—	24.8	6.06	3.94	9.15	0.935	0.221	0.0275	−93	−134	+41	−143	+50
» 11 »	29/9 1959	—	20	6.12	3.91	9.15	0.942	0.205	0.0242	−91	−138	+47	−141	+50
Bh 17 (Pyatigorsk) Morning	22/8 1959	2.5	20	7.04	1.16	8.32	0.118	0.222	0.21	−100	−166	+66	−202	+102
Evening		2.5	21	6.81	5.36	8.32	0.73	0.84	0.492	−120	−176	+56	−193	+73
Bh 1 (Pyatigorsk)	20/8 1959	15	22.6	6.35	0.445	8.45	0.092	0.039	0.0086	−54	−123	+69	−154	+100
» 2 (Matsesta)	1938	—	23.9	6.85	123.9	0.07	15.0	21.4	15.6	−178	−217	+39	−232	+54
» 3 »	1938	—	26.2	6.8	145.9	0.06	17.7	25.1	17.3	−176	−216	+40	−233	+57
» 4 »	1938	—	26.8	6.7	260.3	0.05	29.6	46.8	22.3	−185	−218	+33	−230	+45
» 5 »	1938	—	23.2	6.75	181.4	0.04	20.8	32.4	15.4	−176	−216	+42	−227	+51
» 7 »	1938	—	28.2	7.00	73.3	0.03	6.6	14.9	18.7	−181	−216	+35	−249	+73
» 14 »	1938	—	23	6.95	120.2	0.05	13.0	22.3	19.9	−179	−221	+42	−239	+60
» 17 (Agura)	1938	—	—	6.95	85.8	0.03	8.9	16.3	16.0	−169	−216	+47	−242	+73

In the first case, the dependence of the rate of change in Eh on a_{S^-} at pH = const and $a_{SO_4''}$ = const is determined by

$$\left(\frac{\partial Eh}{\partial a_{S''}}\right)_{pH,\,a_{SO_4}} = -\frac{0.4343 \cdot 0.0074}{a_{S''}}$$

and in the second case by

$$\frac{\partial Eh}{\partial a_{S''}} = \frac{0.4343 \cdot 0.0295}{a_{S''}}$$

The ratio of these values is a constant

$$\frac{(\partial Eh/\partial a_{S^-})}{(\partial Eh/\partial a_{S''})_{pH,\,a_{SO_4}}} = 4.00$$

This means that for any particular error in determining a_{S^-}, the error in computing theoretical Eh for the oxidation of sulfide ion to sulfur may allowably be four times that for the oxidation of sulfide ion to sulfate.

Therefore, as a first approximation, it may be stated that with a particular error in determining $a_{S''}$ the results prove to be indeterminate if

$$\frac{\Delta Eh_{S^-/S^0}}{\Delta Eh_{S''/SO_4}} = 4$$

When

$$\frac{\Delta Eh_{S''/S^0}}{\Delta Eh_{S''/SO_4}} > 4$$

the probability is increased that Eq. (36) is valid. When

$$\frac{\Delta Eh_{S''/S^0}}{\Delta Eh_{S''/SO_4}} < 4$$

on the other hand, it is more likely that the oxidation of hydrogen sulfide is described by Eq. (37). As seen from Table 7, the ratio of errors is always less than unity, which makes it possible to discard Eq. (36) completely. This conclusion leads to negation of the diagrams proposed by Krumbein and Garrels [1952], constructed on formal premises and, consequently, improperly reflecting the conditions under which sulfides form in recent sediments and sedimentary rocks. However, this still does not prove the possibility of using Eq. (37).

As we have pointed out, Zavodnov came to precisely this conclusion, but the errors presented in Table 7 do not instill special confidence.

Under the influence of such data and in light of the investigations of Skopintsev [Skopintsev et al., 1961], the author in due time proposed the idea that the activity of the sulfide ion, depending on the oxidation−reduction potential at low activities, is

most likely determined by some intermediate reaction between

$$S'' \rightleftarrows S^0 + 2e \quad \text{and} \quad S'' + 4H_2O \rightleftarrows SO_4' + 8H^+ + 8e$$

with a marked shift toward the first reaction [Stashchuk, 1963; Stashchuk et al., 1964].

With regret we must accept the fact that this position is incorrect. A series of new experiments and critical examination of data published by different investigators have compelled the author to accept finally the view that sulfide equilibrium is either determined by system (37) or that the sulfur system does not, in general, determine potential. Since this conclusion is extremely important, because the nature of the plotted diagrams will depend on it, let us analyze in more detail some data compelling the author to take up this position.

The examination will begin with the calculation of Table 8. Zavodnov prepared this table on the basis of the fact that $E^0_{S''/S^0} = -0.476$. This value of the standard potential was taken from the work of Maronny and Valensi [1957], who computed it on the basis of the activity of the sulfide ion of system (37) with a correction of formation of the polysulfides and acceptance of data on the hydrogen-sulfide ion from the handbook of Latimer [1952].

Zavodnov borrowed from these investigators only the value of the standard potential. The value of the apparent dissociation constant for hydrogen sulfide, coordinated with the standard potential by Maronny and Valensi, was adopted differently, thus: the second constant was derived from Zavodnov's experimental data, without consideration of polysulfide character, but the first was taken from Golovin [1959]. This led to disagreement among the values and introduced a certain error during computations of the oxidation−reduction potential. We shall try to correct the inaccuracy admitted by Zavodnov, but we shall first make use of some concepts necessary as a basis for what follows.

As is well known, the activity of any ion is determined by its concentration multiplied by the activity coefficient. For the hydrogen-sulfide system we may write the following three equations:

$$[H_2S] + [HS'] + [S] = \sum S$$
$$\frac{[H^+] \cdot \gamma_{H^+} \cdot [HS'] \cdot \gamma_{HS'}}{[HS] \cdot \gamma_{H_2S}} = K_1$$
$$\frac{[H^+] \cdot \gamma_{H^+} \cdot [S''] \gamma_{S''}}{[HS'] \cdot \gamma_{HS'}} = K_2$$

where the symbols in brackets designate molal concentration, and γ the activity coefficient of the corresponding ion. Similarly, as done earlier, we solve this system relative to $[H_2S]$, $[HS']$ and $[S'']$:

TABLE 8. Activity of the Sulfide Ion and Standard Potential
of the Hydrogen–Sulfide–Sulfur System from Corrected Data

lg $\sum a_S$	$a_{S''} \cdot 10^{14}$		Eh, computed		ΔEh		E^0, computed	
	according to Zavodnov	corrected	according to Zavodnov	corrected	according to Zavodnov	corrected	according to Zavodnov	corrected
−3.61	3.68	0.3124	−144	−137	+6	−1	−470	−477
−5.89	0.036	0.0029	−78	−77	+3	+2	−473	−474
−1.98	125.0	8.969	−203	−180	+8	−15	−468	−491
−3.72	7.36	0.6112	−153	−145	+20	+12	−456	−464
−3.71	5.25	0.4232	−148	−140	+15	+7	−461	−469
−3.22	11.60	0.7835	−183	−148	+21	+14	−455	−462
−5.33	0.0102	0.0006	−95	−57	+5	−33	−471	−509
−3.53	6.52	0.5157	−153	−143	+31	+21	−445	−455
−5.92	0.0034	0.0002	−74	−44	+31	+1	−445	−475
−5.93	0.0039	0.0003	−75	−45	+35	+5	−441	−471
−3.95	0.275	0.0167	−134	−99	+41	+6	−435	−470
−3.71	0.242	0.0130	−138	−96	+47	+5	−429	−471
−4.53	2.10	0.1482	−166	−127	+66	+27	−410	−449
−3.85	4.92	0.3427	−176	−138	+56	+18	−420	−458
−4.90	0.086	0.0052	−123	−84	+69	+30	−407	−446
−2.49	156.0	11.181	−217	−182	+39	+4	−437	−472
−2.42	173.0	13.314	−216	−185	+40	+9	−436	−467
−2.23	223.0	14.478	−218	−186	+33	+1	−443	−475
−2.38	154.0	9.2707	−218	−180	+42	+4	−434	−472
−2.72	187.0	14.935	−216	−186	+35	+6	−431	−470
−2.51	199.0	14.465	−221	−186	+42	+7	−434	−469

$$[H_2S] = \frac{\sum S \cdot \gamma_{H_2S}^{-1} \cdot 10^{-2pH}}{10^{-2pH} + 10^{-pH} \cdot K_1 \cdot \gamma_{HS'}^{-1} + K_1 K_2 \cdot \gamma^{-1}}$$

$$[HS'] = \frac{\sum S \cdot K_1 \cdot \gamma_{HS'}^{-1} \cdot 10^{-2pH}}{10^{-2pH} + 10^{-pH} \cdot K_1 \cdot \gamma_{HS'}^{-1} + K_1 K_2 \cdot \gamma_{S''}^{-1}}$$

$$[S''] = \frac{\sum S \cdot K_1 \cdot K_2 \cdot \gamma_{S''}^{-1}}{10^{-2pH} + 10^{-pH} \cdot K_1 \cdot \gamma_{HS'}^{-1} + K_1 K_2 \cdot \gamma_{S''}^{-1}}$$

With the values of pH given in Table 8, we obtain the inequalities

$$K_1 \cdot K_2 \cdot \gamma_{S''}^{-1} \ll 10^{-2pH}$$

$$K_1 \cdot K_2 \cdot \gamma_{S''}^{-1} \ll 10^{-pH} \cdot K_1 \cdot \gamma_{HS'}^{-1}$$

the value $K_1 \cdot K_2 \cdot \gamma_{S''}^{-1}$ may therefore be neglected. Whence

$$[H_2S] = \frac{\sum S \gamma_{H_2S}^{-1}}{1 + 10^{pH} \cdot K_1 \cdot \gamma_{HS'}^{-1}}$$

$$[HS'] = \frac{K_1 \cdot \sum S \cdot \gamma_{HS'}^{-1}}{10^{-pH} + K_1 \cdot \gamma_{HS'}^{-1}}$$

$$[S''] = \frac{K_1 \cdot K_2 \cdot \sum S \cdot \gamma_{S''}^{-1}}{10^{-2pH} + 10^{-pH} \cdot K_1 \cdot \gamma_{HS'}^{-1}}$$

Correspondingly

$$a_{H_2S} = \frac{\sum S}{1 + 10^{pH} \cdot K_1 \cdot \gamma_{HS'}^{-1}}$$

$$a_{HS'} = \frac{K_1 \sum S}{10^{-pH} + K_1 \cdot \gamma_{HS'}^{-1}}$$

$$a_{S''} = \frac{K_1 \cdot K_2 \cdot \sum S}{10^{-2pH} + 10^{-pH} \cdot K_1 \cdot \gamma_{HS'}^{-1}}$$

Since we have given no data on the mineralization and ionic composition of water in this paper, we may determine $\gamma_{HS'}$ by indirect means:

$$[HS'] = \frac{K_1 \cdot \sum S \cdot \gamma_{HS'}^{-1}}{10^{-pH} + K_1^* \cdot \gamma_{HS'}^{-1}}$$

where K_1^* is the first dissociation constant of hydrogen sulfide as adopted by Zavodnov.

After conversion we find

$$\gamma_{HS'} = \frac{K_1^* (\sum S - [HS'])}{[HS'] \cdot 10^{-pH}}$$

but since, with pH values given in Table 8,

$$\sum S = [HS'] + [H_2S]$$

then

$$\gamma_{HS'} = 10^{pH} \cdot K_1^* \frac{[H_2S]}{[HS']}$$

Therefore

$$a_{H_2S} = \frac{\sum S}{1 + \frac{K_1}{K_1^*} \cdot \frac{[HS']}{[H_2S]}} ; \quad a_{S'} = \frac{\sum S \cdot K_1 \cdot K_2 \cdot 10^{2pH}}{1 + \frac{K_1}{K_1^*} \cdot \frac{[HS']}{[H_2S]}}$$

$$\sum a_S = a_{H_2S} (1 + K \cdot 10^{pH})$$

To determine the corrected activities for Table 7 it is necessary to investigate the change in apparent constant with temperature. Zavodnov [1965] gave the following value as a function of temperature

$$K_1 = (0.063 t + 0.02) \cdot 10^{-7}$$

$$pK_2 = 12.934 - 0.0165 (t - 20), \text{ where } t \text{ is temperature, } °C$$

In assuming that the change in K_1 and K_2 with temperature has this character, as explained by Zavodnov, the constants computed by Maronny and Valensi take on the form

$$K_1 = (0.063\,t - 0.577) \cdot 10^{-7}$$

$$pK_2 = 13.973 - 0.0165\,(t - 20)$$

or, in a different form

$$K_2 = 10^{0.0165t - 14.303}$$

whence, in final form,

$$a_{H_2S} = \frac{[H_2S] + [HS']}{1 + \dfrac{(6.3t - 57.7)}{(6.3t + 2)} \cdot \dfrac{[HS']}{[H_2S]}}$$

$$a_{HS'} = \frac{(6.3t - 57.7) \cdot 10^{pH-9}\,([H_2S] + [HS'])}{1 + \dfrac{(6.3t - 57.7)}{(6.3t + 2)} \cdot \dfrac{[HS']}{[H_2S]}}$$

$$\sum a_{S''} = \frac{(6.3t - 57.7) \cdot 10^{0.016t + 2pH - 23.303}\,([H_2S] + [HS'])}{1 + \dfrac{(6.3t - 57.7)}{(6.3t + 2)} \cdot \dfrac{[HS']}{[H_2S]}}$$

$$\sum a_S = ([H_2S] + [HS']\,\frac{1 + (6.3t - 57.7) \cdot 10^{pH-9}}{1 + \dfrac{(6.3t - 57.7)}{(6.3t + 2)} \cdot \dfrac{[HS']}{[H_2S]}}$$

Table 8 shows the data of Zavodnov as compared with corrected data. As we may see from this table, the deviation in computed standard potential is considerable where the dissociation constants of hydrogen sulfide disagree with them, but the deviation is not so great where there is agreement. A more important fact is that the computed values for (37) deviate in both positive and negative directions relative to measured values. This is due to the general character of the distribution of measuring errors. After making a rather simple statistical computation, it is possible to come to a still more important conclusion. As such computation has shown for E^0 computed for experimental data, $\sigma_{Zavodnov} = 17.92$, whereas $\sigma_{calc.} = 13.40$. From a comparison of the dispersions it becomes clear that with the proper choice of constants, not only does the divergence from standard potential diminish but the points are grouped closer to their mean value. It is apparent that sufficiently precise measurements and properly chosen constants may lead to almost complete agreement between computed and actual values.

Such detailed analysis of the given example serves two purposes: first, it shows that the references of geologists to the lack of agreement between actual and computed data, and also the referral of all deviations to the complexities of natural processes, are probably mostly without basis. The given concrete example shows that the use of accu-

rate computations permits us to take the position, with assurance and certainty, that the oxidation of hydrogen sulfide, both in the laboratory and under natural conditions, strictly obeys the reversible reaction

$$H_2S \rightleftarrows S^0 + 2H^{\cdot} + 2e \quad (\text{or} \quad S'' \rightleftarrows S^0 + 2e)$$

Consequently, the oxidation−reduction potential of the hydrogen-sulfide−sulfur system completely determines the activity of the sulfide ions.

Second, by this example we wish again to emphasize the fact that small deviations from initial parameters may cause great deviations in computed values and may even lead to invalid conclusions.

The data of Zavodnov embrace a wide range of water types with extremely variable conditions. As seen from Table 8, the waters are characterized by changes in total activity of hydrogen sulfide from $1.01 \cdot 10^{-2}$ to $1 \cdot 10^{-6}$. Activity of the sulfide ion ranges from $1.5 \cdot 10^{-10}$ to $3 \cdot 10^{-15}$ mole/liter. On these waters, however, oxygen of the air has no effect, being absent entirely or present in only small quantities. In any case, the hydrogen-sulfide system determines the potential.

It is important to take into account the effect a large amount of oxygen will have on $a_{S''}$. Apart from investigations of Skopintsev [Skopintsev et al., 1961], in which the kinetics of the process has been examined, the data of Levchenko are of great interest to us, since here the change in oxidation−reduction potential is due to the effect of oxygen. Levchenko [1950] furnishes results of computing apparent standard potential for hydrogen-sulfide waters held for a long time in contact with atmospheric oxygen. As appears from his descriptions, the experiments were carried out at 20°C, but the dissociation constants used were computed for 25°C. It is not possible to recompute these constants to relate to those adopted in the present work and to consider the error in going from sulfide-ion concentration to activity by using only the author's paper, since there are no data except for $\gamma_{HS'}$, $\gamma_{S''}$, Eh_{meas}, and $0.029 \log [S'']$.

The second part of this interesting experiment was published in a paper by Levchenko and Makarova [1950]. From the data of interest to us, we find here the time of contact with air, Eh, $[H_2S]$, and $[HS']$ but not the value of $\gamma_{HS'}$ or the ionic strength, by means of which it would be possible to compute $a_{S''}$. By combining the tables of the indicated papers into a single one, it is possible to compute the standard potential, as was done for Zavodnov's data, and to examine the effect of oxygen on the change in this potential. We may determine pH in the following way:

$$[HS'] = \sum S \frac{K_1 \cdot \gamma_{HS'}^{-1}}{a_{H^\cdot} + K_{1/\gamma_{HS'}}} \quad (\text{for} \quad pH \leqslant 9)$$

Whence, taking K_1 from Levchenko, we find

$$pH = 7.05 - \lg \frac{[H_2S]}{[HS']}$$

Substituting the values of pH in the equations of activities, we find

$$a_{H_2S} = \frac{\sum S}{1 + 10^{0.106 - \lg \frac{[H_2S]}{[HS']}}}$$

$$a_{HS'} = a_{H_2S} \cdot K_1 \cdot 10^{pH}$$

$$a_{S''} = \frac{\sum S \cdot 9.5 \cdot 10^{-8 - \lg \frac{[H_2S]}{[HS']}}}{10^{\lg \frac{[H_2S]}{[HS']}} + 1.248}$$

The results of Levchenko's experiment and the data we were able to recalculate from his results are given in Table 9. The recomputed data of Zavodnov and Levchenko are shown in Fig. 35. As we may see from this figure, the effect of high oxygen content begins to appear at $\Sigma a_S = 7.25 \cdot 10^{-4}$ mole/liter. In this case, with a sharp rise in Eh, the decline of $a_{S''}$ takes place very slowly. In particular, when Eh = 0.080 V, $a_{S''}$ exceeds the necessary quantity by a factor of $7 \cdot 10^3$. When Eh = +0.339 V, the factor is $6 \cdot 10^{17}$, and at Eh = +0.382 V it is 10^{19}.

When $\Sigma a_S > 7 \cdot 10^{-4}$ mole/liter, the effect of oxygen is not appreciable, and, despite the fact that, along with the oxidation of hydrogen sulfide, not only elemental sulfur appears but also other reaction products (thiosulfate, sulfite, sulfate), the activity of

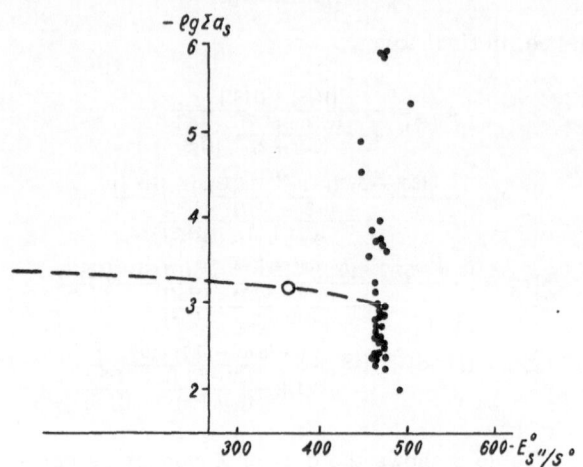

Fig. 35. Standard potential of the sulfide-ion – sulfur system, computed from analytical data of mineral waters.

TABLE 9. Oxidation–Reduction State of the Sulfide Ion Found in Contact with Atmospheric Oxygen

Duration of oxidation	E^0, V	$H_2S \cdot 10^3$, moles per liter	$HS' \cdot 10^3$, g-ions per liter	$S_2O_3'' \cdot 10^3$, g-ions per liter	$SO_3'' \cdot 10^3$, g-ions per liter	$SO_4'' \cdot 10^3$, g-ions per liter	$S \cdot 10^3$, g-atoms per liter	$\Sigma a_S \cdot 10^3$, g-ions per liter	$\lg \sum a_S$	$a_{S''} \cdot 10^{10}$, g-ions per liter	E^0, V
0	−0.175	4.00	2.89	0.005	0.001	0.025	0.016	5.400	−2.268	1.79	−0.463
3 hours	−0.175	3.08	2.81	0.02	0.003	0.025	0.016	4.484	−2.348	2.177	−0.460
6 hours	−0.175	2.83	2.91	0.025	0.003	0.025	0.03	4.522	−2.34	2.524	−0.458
12 »	−0.177	2.73	2.81	0.035	0.003	0.045	0.08	4.365	−2.36	2.443	−0.461
24 »	−0.184	1.44	2.64	0.055	0.025	0.045	0.22	3.209	−2.49	1.882	−0.471
36 »	−0.187	1.22	2.49	0.055	0.003	0.040	0.28	2.714	−2.57	4.143	−0.461
2 days	−0.188	1.20	2.45	0.075	0.003	0.045	0.37	2.670	−2.57	4.076	−0.465
3 »	−0.186	1.01	2.41	0.11	0.003	0.06	0.88	2.464	−2.61	4.654	−0.461
4 »	−0.192	0.52	2.09	0.12	0.003	0.06	0.92	1.801	−2.74	6.655	−0.463
5 »	−0.191	0.43	1.97	0.15	0.003	0.065	1.05	1.641	−2.78	7.125	−0.461
6 »	−0.190	0.45	1.73	0.175	0.005	0.075	1.37	1.510	−2.82	5.282	−0.464
7 »	−0.190	0.51	1.51	0.17	0.005	0.08	1.39	1.428	−2.85	3.579	−0.469
9 »	−0.188	0.50	1.34	0.22	0.005	0.09	1.48	1.260	−2.90	2.721	−0.470
12 »	−0.190	0.40	1.21	0.28	0.005	0.11	2.02	1.136	−2.94	2.934	−0.471
15 »	−0.186	0.33	1.19	0.25	0.003	0.13	2.09	1.010	−2.99	3.742	−0.464
19 »	−0.179	0.23	0.90	0.33	0.005	0.16	2.35	0.782	−3.12	2.798	−0.461
25 »	−0.080	0.21	0.84	0.28	0.003	0.18	2.45	0.725	−3.14	2.663	−0.362
33 »	+0.339	0.08	0.39	0.28	0.003	0.57	2.60	0.321	−3.49	1.498	+0.049
43 »	+0.382	0.02	0.05	0.005	Tr	1.92	1.75	0.050	−4.30	1.009	+0.087
58 »	+0.660	None	None	None	None	2.24	1.57	—	—	—	—
73 »	+0.480	»	»	»	»	2.92	0.56	—	—	—	—
123 »	+0.440	»	»	»	»	3.20	0.10	—	—	—	—
4 months	—	»	—	—	—	—	—	—	—	—	—

the sulfide ion is determined strictly by the oxidation−reduction potential of system (37). In the absence of oxygen, or when the content is low, the activity of the sulfide ion, down to very low concentrations, sufficient for plotting stability fields of the sulfides, may be computed from system (37).

The effect of oxygen on change in potential is also pointed out in the investigation of Skopintsev [1957] on the Black Sea. In referring to this material as qualitative, since the necessary data for computation are lacking, we note that the effect of oxygen begins to appear at an H_2S content of approximately $(0.5-1.0) \cdot 10^{-4}$ mole/liter. Naturally, whenever the time for this arrives, the hydrogen-sulfide system ceases to determine the potential. But, it is still not clear to what extent this time can be accurately determined or how the pattern of changes in sulfide ion under these conditions can be explained.

It seems to us that thorough investigations along this line are necessary, since only they will allow us to find a more complete answer to the question concerning the conditions under which sulfides form. While it is necessary to restrict ourselves to these facts, we should keep in mind that the conditions under which iron sulfides form may be investigated from two positions.

1. The oxygen-free (or low oxygen-content) system with analytically determinable contents of hydrogen sulfide. In this case we have an equilibrium process in which the activity of the sulfide ion may be computed from Eq. (37).

2. The system of mixed hydrogen sulfide and oxygen. In this case, we have an nonequilibrium process. The sulfide ion here at a definite stage, because of the kinetic peculiarities, is not reflected by system (37). Its activity is therefore greater than suggested by the indicated system. This aspect is made use of by sulfur bacteria for manifestation of their activities. This factor is also responsible for the fact that the activity of the sulfide ion cannot always be strictly calculated from Eq. (37).

Mineral Forms of Iron Sulfides in Sedimentary Rocks and Recent Sediments

Earthy varieties of iron sulfides, especially those that are almost always disseminated through the surrounding rocks, are very difficult to identify. That is why most geologists at the present time describe the sulfides in sedimentary rocks and recent

sediments by means of the terms pyrite, marcasite, hydrotroilite, and melnikovite. It is true that recently we have begun to find even pyrrhotite rather frequently [Chukhrov, 1936; Ivanov, 1957; Bobrovnik, 1958; Stashchuk et al., 1964].

Whereas the first two sulfides give beautiful crystal forms and occur in individual aggregates, this cannot be said for the others. The terms melnikovite, hydrotroilite, and even pyrrhotite, representing material found in sedimentary rocks, are most probably general concepts, not reflecting precise representations of structure or chemical composition of these sulfides. However, since we can construct diagrams on the basis, as a minimum, of precise chemical composition of a compound, it is necessary to examine what is now involved in the concepts of such sulfides as hydrotroilite, melnikovite, and pyrrhotite.

After the investigations of Braconnt [1832] most investigators adopted the view that the black color of mud is due to the presence of iron sulfide. In 1901 Sidorenko gave this compound the name hydrotroilite in accordance with the idea that $FeS \cdot H_2O$ was part of the concept of this sulfide. However, whether the black sooty material widespread in muds actually has this composition is not yet clear. Doelter [1926] gave it the formula $FeS \cdot nH_2O$. Dal'nichenko and Chigirin [1926] represented it by $Fe(HS)_2$. Data on hydrotroilite have been obtained from studies of recent sediments, but, as Lepp has validly remarked, hydrotroilite "has never been studied mineralogically and there is some question that it is actually a variety" [Lepp, 1957].

Black, outwardly amorphous material similar in form to hydrotroilite of recent sediments was discovered in 1906 by Doss [1911] in Miocene clays of the Melnikov farm, and he gave it the name melnikovite. The chemical composition of this sulfide from Bh 2 is expressed by the formula Fe_5S_7, and from Bh 3 by FeS_2. Doss believed that the most suitable conditions for the formation of melnikovite were those found on the floor of the Black Sea. Subsequent discoveries of such material have led to examinations of a great variety in regard to the Fe/S ratio.

It is true that investigators most commonly seek to identify melnikovite with the formula FeS_2, plus an accompanying arbitrary quantity of FeS. Ehrenberg [1928], for example, represented melnikovite in this way. The idea has been revived recently by Volkov [1964]. Lindren [1926] gave melnikovite the formula FeS, which identifies it completely with the hydrotroilite of Sidorenko. Angel and Scharizer [1952] stated that melnikovite and hydrotroilite are one and the same mineral. It has not been possible to determine the precise chemical composition of this material, however.

In the twenties and thirties, structural analysis had not yet been widely used in mineralogy, and it was therefore impossible to state precisely what melnikovite or hydrotroilite was: an independent mineral species or a mixture of minerals. Therefore, along with descriptions of new discoveries of these minerals, papers have consistently appeared to raise doubts concerning the existence of melnikovite as a mineral species. For example, Berz [1922] and Behrend [Behrend and Berg, 1927] categorically deny the existence of melnikovite. Rodt [1916] doubted that the compound dealt with by Doss is characterized by the formula FeS_2. This same view was maintained by Chukhrov.

The term melnikovite began to be used for earthy varieties of marcasite or pyrite [Ramdohr, 1950-60]. Essentially, before the appearance of the structural investigations of Lepp [1957], it was impossible to state reliably what melnikovite was. Unfortunately, the structural investigations of the compound synthesized by Lepp were not accompanied by chemical analysis. Lepp therefore gave no definite formula for melnikovite, indicating merely that the S/Fe ratio is greater than necessary for FeS but less than in FeS_2. Nevertheless, on the basis of this work, from a comparison of x-ray patterns, Volkov and Ostroumov [1957] concluded that melnikovite exists as a type of FeS_2 in recent Black Sea sediments. It was soon shown, however, that this incautious conclusion was erroneous. A study of x-ray structural data and qualitative reactions on $Fe^{...}$ from melnikovite concretions of Black Sea deposits, carried out by Volkov, Polushkina and Sidorenko [1963], came to the following conclusion: the melnikovite structure exhibits dense packing of S" ions with the tetrahedral and octrahedral sites being occupied by divalent and tervalent iron. The ratio of filled tetrahedral sites depends on the $Fe^{..}/Fe^{...}$ ratio in the sample. When the ratio diminishes, the total number of iron atoms and the Fe/S ratio will decline, and the composition of the mineral will diverge somewhat from the stoichiometric formula Fe_3S_4.

Thus, the mineral called melnikovite and synthesized by Lepp (since the x-ray patterns cited by Lepp, Volkov, Polushkina, and Sidorenko agree in all characteristic features given by Doss) has the structure of magnetite, shows a qualitative reaction to the ferric ion, and in chemical relations, as magnetite is described by the formula $FeO \cdot Fe_2O_3$, may be described by $FeS \cdot Fe_2S_3$. Such a mineral

was also found by these investigators in the Oligocene clays of Mangyshlak.

A brief note by Polushkina and Sidorenko [1963] opened up new possibilities in the study of iron sulfides in recent sediments and sedimentary rocks. Although chemists were obtaining Fe_2S_3 from cold solutions as early as the end of the last century, and the formation of this compound at the present time by reaction between ferric hydroxide and hydrogen sulfide is widely accepted [Kuznetsov and Sagalovskii, 1954] even though the cubic sulfide Fe_3S_4 was synthesized by Yamaguchi and Katsurai as early as 1960, still the fact described by Polushkina and Sidorenko indicates that Fe_3S_4, in which the theoretical ratio of $Fe^{..}/Fe^{...}$ is 1/2, forms under natural conditions as well as by artificial means, even with high contamination of hydrogen sulfide, i.e., under conditions known to be reducing.

Outwardly the melnikovite concretions discovered by Volkov [1961b, 1964] are characterized by different forms and are black or dull black. Similar dull black aggregates have been found under conditions of strong hydrogen-sulfide contamination in Sivash muds and have been identified as pyrrhotite [Stashchuk et al., 1964]. It may be thought that the dark black balls described by Rauzer-Chernousova [1928] for the muds of a bay in Krugloi Lake also consist of sulfides of ferric iron. Ostroumov [1953b] noted similar aggregates in muds of the Black Sea and called them "transitional forms between colloidal black masses of hydrotroilite and iron sulfide." As seen from this list, cubic iron sulfide (and perhaps pyrrhotite) must be rather widespread in recent sediments.

In 1956 Erd [Erd et al., 1956] discovered, and in 1957 described, a trigonal variety of sulfide having the formula Fe_3S_4, found in sedimentary rocks and called smythite. At temperatures above 40°C smythite decomposes; consequently, it can form only at low temperatures. It is similar to pyrrhotite in chemical properties and composition.

Skinner [Skinner et al., 1964] furnished a detailed description of a sulfide (gregite) discovered in Tertiary clays of California (U.S.A.). The sulfide has an ideal spinel structure and is expressed by the formula Fe_3S_4. As explained below, the melnikovite described by Polushkina and Sidorenko from muds of the Black Sea [1963] and from the Melnikov farm [1968] and the iron sulfide we discovered in muds of the Sivash and in Maikopian deposits of the Crimea [Stashchuk and Kropačeva, 1969] are one and the same mineral. The International Mineralogical Association has assigned it the name gregite [Bonshtedt-Kupletskaya, 1970]. In analyzing discoveries of the various sulfides discovered in sedimentary rocks recently, Chukhrov [Chukhrov et al., 1965] noted the diversity of iron sulfides found in nature (pyrite, marcasite, troilite, pyrrhotite, melnikovite, gregite, and smythite) representing the Fe−S system, differing from each other in composition or in structure with identical composition.

What was responsible for the opinion maintained by investigators until recent years that only sulfides of ferrous iron could be found in recent sediments and sedimentary rocks? The principal reason for this is that it has been impossible to analyze sulfides with different forms of iron and sulfur. In particular, in regard to determining ferrous iron in the variety of chemical analytical methods used in laboratories of the Geochemical Division of the Central Scientific-Research Geological-Exploration Institute (TsNIGRI), an unequivocal instruction is given: "It is important to determine ferrous iron in sulfide ores decomposed by acid" [Knipovich, 1936, p. 40].

In the handbook "Analysis of Mineral Raw Materials" for 1956, the following is stated in reference to iron ores decomposed by acid: "this determination (ot terrous iron) is impossible if in the material being analyzed there are strong oxidizing agents (such as the mineral oxides of tervalent and quadrivalent manganese) or reducing agents (large quantities of sulfides or organic matter)." The linkage of hydrogen sulfide also proves to have little effect. Such methods in the "presence of very easily decomposed sulfides give no positive results" [Analysis of Mineral Raw Materials]. As Stukalova [1947] pointed out, experiments with the addition of mercury salts for determining forms of iron in silicates with even a small admixture of pyrrhotite do not give reliable results, and the introduction of copper sulfate into a pyrrhotite sample leads to different results depending on length of time the sample is treated. From the thermodynamic viewpoint, this is rather easily explained. For example, with iron sulfide of the type $FeS \cdot Fe_2S_3 = (Fe_3S_4)$, the free energy of which is 88.74 kcal, the following reaction takes place in an acid environment:

$$Fe_3S_4 + 6H^{\cdot} = 3Fe^{\cdot\cdot} + S + 3H_2S$$
$$\Delta G^0_{reaction} = 8.22, \quad -\lg K = 6.03$$

whence we may calculate

$$a_{Fe^{\cdot\cdot}} = \frac{1}{10^{2\,(pH-1)} \cdot a_{H_2S}}$$

Thus, the more quickly free hydrogen sulfide is removed from solution, being separated during

breakdown of sulfide in acid, and the lower the pH of the decomposition, the more ferrous ion there proves to be at the expense of the ferric ion. The content of this ion is determined by two independent variables, and this means that, depending on what hydrogen sulfide is bound to (salts of copper or mercuric chloride), the concentration and volume of added acid and the kinetics of the process will determine the content of the ferrous ion.

As pointed out by Kumai [1958], even with the compound Fe_2S_3 in acid solutions, we observe the reaction

$$Fe_2S_3 + 4H^{\cdot} = 2Fe^{\cdot\cdot} + S + 2H_2S$$

which naturally allows no possibility of making accurate analysis of the form of iron in sulfides readily decomposed by acid. Therefore, when analyzing sulfides we generally determine total iron and total sulfur, which are then recomputed into a sulfide.

By this method completely different minerals may give the same results, and vice versa. In particular, Fe_3S_4 may be interpreted either as $FeS \cdot Fe_2S_3$ or $2FeS \cdot FeS_2$. The way the formula is written depends entirely on the viewpoint maintained by the investigator. If the view is held that in the muds "processes of reduction involve primarily quadrivalent manganese and then tervalent iron and its mobile form (type of hydroxide), and only after this do sulfates begin to be reduced" [Ostroumov and Fomina, 1959, p. 385], it is then logical to write the formula $2FeS \cdot FeS_2$.

Since this view is dominant in geology at the present time [Strakhov, 1956a, 1960b], it is perfectly natural that many investigators associate iron sulfides forming in sediments only with the ferrous iron, despite that fact that in the laboratory chemists have long obtained sulfides of different types, from Fe_2S_3 to FeS.

Recently, in view of factual information, opinion concerning a strict sequence in the reduction of elements, corresponding to their standard potentials, has begun to undergo certain changes. In particular, in a study of the geochemistry of iron in the northern part of the Indian Ocean, it was noted that the reduction of sulfates takes place simultaneously with reduction of iron, not after this process [Isaeva, 1964].

The use of accurate structural investigations on natural mineral forms has allowed us to shed some light on sulfides with the ferric ion and has thereby raised doubts concerning the view that the sulfate ion begins to be reduced only after complete reduction of the ferrous ion. To the point, as already discussed, the formation of the sulfide ion by reduction of sulfates by no means presumes invariably a low potential, or, consequently, conditions under which all iron should occur in the reduced form. In the laboratory, under conditions approaching those in nature, the cubic sulfide of iron was synthesized by Berner [1964]. This investigator, after recalculating the x-ray patterns of Volkov [1961b], also concluded that the concretions of dark dull iron sulfide in muds of the Black Sea have a composition not of $nFeS \cdot FeS_2$ but of Fe_3S_4. The results of Berner's experiments are shown in Tables 10, 11, and 12.

The experiments of Berner were repeated by Korolev and Kozerenko [1965], who obtained similar results (Table 13). Differences were the following: Berner obtained pyrrhotite only from metallic iron whereas Korolev and Kozerenko synthesized this compound from solutions of the ferrous ion. We should cite here also the data on synthesis of this compound from ferric hydroxide by Rosenthal [1956].

Unfortunately, neither in the experiments of Berner, nor in the investigations of Erd, Evans, and Richter, nor in the experiments of Korolev and Kozerenko were sufficient data given for explaining the quantitative relations between such parameters as a_{H_2S}, pH, $a_{Fe^{\cdot \cdot}}$, and the type of sulfide that forms. Data on the oxidation—reduction potential given by Korolev and Kozerenko cannot be used, since they reflect the oxidation—reduction state of the initial solution. We may state, merely, that these data correspond to some reversible or steady system with $E^0 \approx 0.200$ V.

One's attention is drawn to the report of Temple [1964] concerning the fact that in experiments of Freke and Tate the sulfide obtained in a medium where sulfate-reducing bacteria were active had the composition $2FeS \cdot Fe_2S_3$. Butlin and Postgate obtained Fe_4S_5 with the reduction of sulfates by sulfate-reducing bacteria at a rate of 203-262 ml/liter·h.

TABLE 10. Experiments with Goethite (after Berner [1964b])

Initial material	Final pH	Temperature of aging, °C	Period and method of aging	Results
Natural geothite (< 2μ) + H_2S	4	20−25	14 h − N	Pyrite + tetragonal Fes + unreacted goethite
Synthetic goethite (aged 2 days) + H_2S	4	20−25	2 days − N	Pyrite + tetragonal FeS
Synthetic goethite (aged 2 days) + acetic acid	7	20−25	2 days − M	Amorphous FeS
Synthetic goethite (aged 2 weeks) + Na_2S + acetic acid	8	20−25	11 days − M	One broad band ≈ 5Å + sulfur
Synthetic goethite (aged 2 weeks) + H_2S + NH_4OH	6.5	20−25	22 days − E	Tetragonal FeS (very broad lines) + sulfur
Synthetic goethite (aged 18 months) + Na_2S + acetic acid	8	40−42	5 days − M	Tetragonal FeS (very broad lines)
Synthetic goethite (aged 18 months) + H_2S + NH_4OH	6	60−65	20 h − E	Tetragonal FeS (very broad lines) + sulfur

Characteristics of iron aging:

E — excess internal pressure;

M — minimal air space above solution;

S — large air space above solution;

O — open system relative to atmosphere; bubbling through a solution of H_2S;

N — experiment in an atmosphere of nitrogen.

TABLE 11. Experiments with Metallic Iron (after Berner [1964b])

Initial material	Final pH	Temperature of aging, °C	Period and method of aging *	Results
Iron wire (reagent grade) + H_2S	4	20−25	18 h − E	Tetragonal FeS
Steel + H_2S	4	20−25	3 h − E	Tetragonal FeS
Steel + H_2S	4	40−42	4 days − M	Tetragonal FeS + pyrrhotite (+ pyrite)
Iron spectrographic electrode + H_2S	4	80−85	8 h − E	Tetragonal FeS + pyrrhotite
Steel + Na_2S + Beckman buffer	7	20−25	1 day − M	Tetragonal FeS + sulfur
The same	9	20−25	3 days − M	Tetragonal FeS
Iron powder reactively pure + Na_2S + Beckman buffer	7	40−42	6 days − M	Tetragonal FeS + unreacted iron

*Symbols for iron aging as in Table 10.

TABLE 12. Experiments with Dissolved Ferrous Ion (after Berner [1964b])

Initial Material	Final pH	Temperature of aging, °C	Period and method of aging *	Results
$FeSO_4$ + H_2S	3	20−25	2 h − E	Amorphous FeS
"Mohr's salt" + H_2S	3	20−25	6 days − M	Tetragonal FeS + cubic iron sulfide + sulfur
$FeSO_4$ + H_2S	3	60−65	2 days − M	Cubic iron sulfide + sulfur (+ tetragonal FeS + pyrite)
	3	85−95	30 min − E	Tetragonal FeS
"Mohr's salt" + H_2S	3	85−95	15 min − E	Tetragonal FeS
$FeSO_4$ + H_2S	3	85−95	15 min − O	Cubic iron sulfate + pyrite + marcasite
"Mohr's salt" + H_2S	3	85−95	15 min − O	The same
$FeSO_4$ + H_2S	3	60−65	2 days − S	Pyrite + marcasite (scum at surface of solution)
$FeSO_4$ + H_2S	3	40−42	25 days − S	Pyrite + marcasite (within sample)
$FeSO_4$ + Na_2S + acetic acid	7	20−25		Amorphous FeS
$FeSO_4$ + Na_2S + Beckman buffer	7	40−42	7 days − M	Tetragonal FeS (+ sulfur)
$FeSO_4$ + Na_2S + acetic acid	7	20−25	5 days − M	Amorphous FeS
$FeSO_4$ + Na_2S + Beckman buffer	9	20−25	7 days − M	Amorphous FeS
$FeSO_4$ + Na_2S + Beckman buffer	9	40−42	200 days − M	Tetragonal FeS + sulfur
$FeSO_4$ + Na_2S + Beckman buffer	9	60−65	2 days − M	Cubic iron sulfide + sulfur + broad band ≈ 5Å
$FeSO_4$ + Na_2S + Beckman buffer	9	60−65	100 days − M	Tetragonal FeS (+ sulfur)

*Symbols for iron aging in Table 10.

TABLE 13. Conditions for Synthesis of Iron Sulfide (after
Korolev and Kozerenko [1965])

Serial No.	Composition of initial solution		pH of initial solution	Eh, mV	t, °C	Composition of resulting precipitate
	Fe, mg per liter	H_2SO_4, ml				
1	1600	0.5	7.5—7.0	—250	150	Pyrrhotite
2	1000	0.5	7.5—7.0	—250	150	»
3	1000	0.9	5.5	—125	150	»
4	1000	1.5	3.0	+20	150	»
5	1200	1.5	3.0	+20	150	Pyrrhotite (admixture of smythite), marcasite
6	1600	1.5	3.0	+20	150	Pyrrhotite (admixture of smythite)
7	800	1.5	3.0	+20	150	Pyrrhotite
8	600	1.5	7.5—7.0	—250	150	Pyrrhotite, melnikovite
9	400	0.5	7.5—7.0	—250	150	»
10	600	0.5	7.5—7.0	—250	240	Pyrrhotite, pyrite
11	400	0.5	7.5—7.0	—250	240	Pyrite, pyrrhotite
12	600	0.7	6.5	—180	150	Melnikovite, pyrrhotite
13	600	0.6	5.5	—125	150	Pyrite, pyrrhotite
14	300	0.5	7.5—7.0	—250	150	Pyrite (admixture of pyrrhotite)
15	200	0.7	6.5	—180	150	Pyrite
16	200	0.9	5.5	—125	150	»
17	200	1	4.5	—70	150	Pyrite (admixture of marcasite)
18	200	1.2	3.6	—10	150	Pyrite, marcasite
19	200	1.2	3.6	—10	90	» »
20	200	1.2	3.6	—10	240	» »
21	200	—	4.5	—70	150	Marcasite (admixture of pyrite)
22	1600	0.5	7.5—7.0	—100	150	Pyrite
23	800	1.5	3.0	+90	150	Pyrite, marcasite
24	200	—	5.5	—125	150	» »

Interesting experiments shedding light on the specific conditions during formation of sulfides were carried out by Kumai [1958]. Considering the fact that the determination of different forms of iron in sulfides cannot be made, this investigator explained the products that formed in his experiment by indirect means, without referring to chemical analysis. The experiments of Kumai involved the following. A vessel for chemical reactions is charged with a suspension with no free oxygen, containing ferric

hydroxide in the amount $2 \cdot 10^{-3}$ mole. Then 150 ml of hydrogen sulfide is blown through the suspension. The following reactions might take place:

$$2Fe(OH)_3 + 3H_2S = Fe_2S_3 + 6H_2O \qquad (43)$$

$$2Fe(OH)_3 + 3H_2S = 2FeS + S + 6H_2O \qquad (44)$$

Thus, the amount of elemental sulfide given off during the reactions should be equivalent to the amount of reduced ferric hydroxide. The experimental results are shown in Table 14.

TABLE 14. Reduction of Ferric Hydroxide under the Influence
of H_2S at Different Values of pH (after Kumai [1958])

Serial number	Fe(OH)$_3 \cdot 10^3$ moles	pH		Amount of extracted sulfur, 10^3 g-atom	Yield of FeS, %	Duration %
		before introduction of H_2S	after introduction of H_2S			
1	2.00	12.50	9.89	0.000	0.0	2
2	2.00	11.75	8.70	0.042	4.2	2
3	2.00	11.70	8.07	0.206	20.6	2
4	2.00	8.50	7.20	0.555	55.5	2
5	2.00	7.00	5.14	0.649	64.9	2
6	2.00	6.00	4.88	0.654	65.4	2
7	2.00	5.00	4.83	0.655	65.5	2
8	2.00	4.00	6.47	0.828	82.8	2
9	2.00	2.98	6.07	0.876	87.6	2
10	2.00	12.3	10.7	0.000	0.0	17
11	2.00	12.3	9.90	0.031	3.1	19
12	2.00	12.3	8.30	0.167	16.7	44
13	2.00	6.01	4.93	0.684	68.4	20
14	2.00	12	11.8	0.000	0.0	—
15	2.00	11.8	8.1	0.327	32.7	—

In accordance with reactions (43) and (44), 1.5 mole of H_2S was consumed for one mole of $Fe(OH)_3$. The amount of hydrogen sulfide introduced at 20°C amounted to

$$\frac{273 \cdot 16 \cdot 150}{293 \cdot 16 \cdot 22 \cdot 416 \cdot 10^3} = 6.24 \cdot 10^{-3} \text{ mole}$$

This indicates that the amount of hydrogen sulfide blown through was 2.08 times the amount necessary. The numbers obtained, however, still give no view of the activity of the still-dissolved hydrogen sulfide and of the hydrogen sulfide undergoing reaction. It is perfectly natural that part of the hydrogen sulfide, not having been dissolved, collected over the solution, and, after equilibrium was established above the reacting mixture, remained there, creating a definite partial pressure of H_2S. The content of hydrogen sulfide in the reactive solution may be calculated more or less accurately by the change in pH.

In the first experiment, as shown by the change in pH, after introduction of hydrogen sulfide the activity of the H' ions increased to $10^{-9.89} - 10^{-12.5}$ mole. This increase was connected with the fact that the excess of hydrogen sulfide having been dissolved in water within the indicated limits of pH decomposes almost entirely to H' and HS' ions. Consequently, increase in number of H' ions must correspond to the increase in number of HS' ions. The activity of HS' ions in solution, not bound with ferric hydroxide, is equal to

$$10^{-9.89} - 10^{-12.5} = 1.288 \cdot 10^{-10} - 3.162 \cdot 10^{-13} = 10^{-9.82}$$

that is, $a_{HS'} = 1.51 \cdot 10^{-10}$ with an allowable error at which $a_{S''}$ may be neglected. On the basis of the value of $a_{HS'}$ obtained and the pH value, we compute the activity of H_2S:

$$\frac{10^{-9.82} \cdot 10^{-9.82}}{10^{-7.166}} = 3.36 \cdot 10^{-13}$$

Recalculating the other experiments in similar fashion, we have prepared a table of all possible parameters on which the chemical composition of the resulting sulfides depend (Table 15). We should note that the indicated calculations are not free of one fundamental error, the removal of which was unfortunately impossible. The fact is that at low values of pH, depending on the content of hydrogen sulfide, the following reaction takes place:

$$2Fe(OH)_3 + H_2S + 4H' = 2Fe'' + S + 6H_2O$$

These reactions are accompanied by alkalization of the environment because of consumption of the H ions. Therefore, at low pH values, results of the experiment may be too low. The dominant value of these reactions lies in the eighth and ninth experiments, where an increase in pH instead of decrease is observed after addition of hydrogen sulfide. This fact attests to the fact that at low pH values the ferrous ion is completely in solution. Therefore, the solid phase under acidic conditions again acquires a composition approaching Fe_2S_3. Indirect proof of this view may be found in the experiments of Berner [1964b] and Korolev and Kozerenko [1965], in which Fe_3S_4 was separated in the solid phase under acidic conditions. However, it is hard to maintain that Fe_3S_4 will always precipitate under acidic conditions. Such a viewpoint requires thorough experimental confirmation.

As becomes clear from the experiments of Kumai, even prolonged maintenance of the suspension after introduction of hydrogen sulfide produces no perceptible change in the Fe''/Fe''' ratio. This leads us to maintain that the system is in a state of equilibrium. Let us evaluate the results of Kumai's experiments from the viewpoint of an equilibrium system, taking into account the fact, of course, that some parameters are approximate because of the statements made above.

TABLE 15. Computed Activity of Free Residual Hydrogen Sulfide in the Experiments of Kumai

pH		$a_{HS'}$	a_{H_2S}	$\sum a_S$	Duration, h	$\lg \sum a_S$
before introduction of H_2S	after introduction of H_2S					
12.5	9.89	$1.51 \cdot 10^{-10}$	$3.36 \cdot 10^{-13}$	$1.52 \cdot 10^{-10}$	2	−9.90
11.75	8.70	$2.00 \cdot 10^{-9}$	$5.82 \cdot 10^{-11}$	$2.06 \cdot 10^{-9}$	2	−8.69
11.70	8.07	$8.51 \cdot 10^{-9}$	$8.43 \cdot 10^{-10}$	$9.35 \cdot 10^{-9}$	2	−8.03
8.50	7.20	$5.99 \cdot 10^{-8}$	$5.26 \cdot 10^{-8}$	$1.12 \cdot 10^{-7}$	2	−6.95
7.00	5.14	$7.14 \cdot 10^{-6}$	$7.46 \cdot 10^{-4}$	$7.53 \cdot 10^{-4}$	2	−3.12
6.00	4.88	$1.22 \cdot 10^{-5}$	$1.61 \cdot 10^{-3}$	$1.62 \cdot 10^{-3}$	2	−2.79
12.3	10.7	$1.94 \cdot 10^{-11}$	$3.88 \cdot 10^{-15}$	$1.94 \cdot 10^{-11}$	17	−10.71
12.3	9.90	$1.25 \cdot 10^{-10}$	$1.58 \cdot 10^{-13}$	$1.25 \cdot 10^{-10}$	19	−9.90
12.3	8.30	$5.01 \cdot 10^{-9}$	$2.51 \cdot 10^{-10}$	$5.26 \cdot 10^{-9}$	44	−8.28
6.01	4.93	$1.08 \cdot 10^{-5}$	$1.27 \cdot 10^{-3}$	$1.28 \cdot 10^{-3}$	20	−2.89
12	11.8	$5.85 \cdot 10^{-13}$	$9.30 \cdot 10^{-18}$	$5.85 \cdot 10^{-13}$	—	−12.23
11.8	8.1	$7.94 \cdot 10^{-9}$	$6.3 \cdot 10^{-10}$	$8.57 \cdot 10^{-9}$	—	−8.07

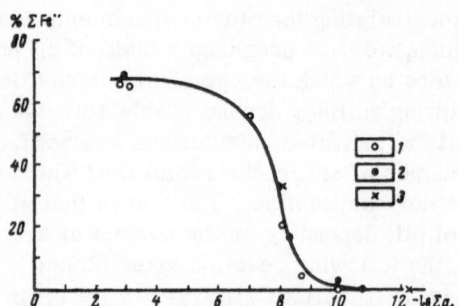

Fig. 36. Relation of log Σa_S to amount of reduced ferric hydroxide (from experimental results of Kumai [1958]). 1) Duration of experiment, 2 hours; 2) duration of experiment, many hours; 3) subsequent addition of hydrogen sulfide.

As seen from Table 14, the pH value in principle reflects the number of ferric ions reduced, but this relationship is not always well defined. In addition, a difficulty arises in making this comparison because it is not known what value of pH is necessary for comparing the results: final or initial. It proves to be clearest and simplest to obtain the relationship between the amount of reduced ferric ion and the total dissolved hydrogen sulfide remaining after the reaction (Fig. 36). This relationship permits us to make the following conclusion: the nature of the reduction of ferric hydroxide is determined by the activity of the hydrogen sulfide. Kumai himself drew the specific conclusion that Fe_2S_3 goes to FeS with decline in pH, whereas the breakdown of Fe_2S_3 to FeS and S begins at pH 8. Volkov [1964] uses this as proof that in marine muds, where pH < 8, there is no possibility of sulfides forming with the ferric ion. It seems to us that he somewhat improperly interprets the results of the experiment.

Let us examine the proof of Kumai. Hydrogen sulfide was added to a suspension of ferric hydroxide containing $2 \cdot 10^{-3}$ mole at a pH $\approx$ 12, thereby lowering the pH to 11.8. Sulfur was not detected during the extraction. After this, hydrogen sulfide was again added, and the pH was thereby lowered to 8.1. During the subsequent extraction, $0.327 \cdot 10^{-3}$ g-atom of sulfur was determined. This served as the basis for concluding that the breakdown of Fe_2S_3 begins at pH 8. One is struck by the fact that the pH value was lowered by hydrogen sulfide; i.e., the activity of H_2S increased in the reaction mixture. Recalculation of the data in Tables 14 and 15 shows that the points of the experiment lie on a curve that reflects a general relationship between Σa_S and the yield of FeS. This experiment convinces one

that the degree of reduction of ferric hydroxide is in great part determined by the activity of hydrogen sulfide and that, in the given case, there is no meaning in speaking of the breakdown of Fe_2S_3. In any case, it is impossible to state that all the Fe_2S_3 goes to FeS.

As has been pointed out, Kumai believes that change in pH is a more important factor in the reduction of iron. Since the experiments were of an applied character — made for the purpose of investigating the possibility of using ferric hydroxide as an adsorbent of hydrogen sulfide during industrial desulfurization [Kumai, 1957] — this investigator did not consider it necessary to check his proposed use of another acid not combined with the S" ion. Therefore, the question of whether the pH value or a_{H_2S} has the greater effect on the degree of reduction of ferric hydroxide is still open.

The following conclusions may therefore be drawn, upon which the computations will be based.

1. The view that in sedimentary rocks iron sulfides are represented only by pyrite, marcasite, and hydrotroilite of the type $FeS \cdot nH_2O$ or melnikovite of the type FeS_2 is far from absolute, and in the case of melnikovite it is inaccurate. As experimental data and actual geological studies have shown, in recent and ancient sediments sulfides are represented by minerals having variable quantities of the ferric and ferrous ions, i.e., sulfides of the type pyrrhotite, smythite, and gregite, and possibly other forms. Up till the present, discoveries of such sulfides have been considered singular because of the disperse character of the minerals, the difficulty of separating them in pure form, and the absence of methods permitting reliable determination of the forms of sulfur and iron present in the given sulfide.

2. Analysis of experimental data indicates that the conditions under which sulfides with different $Fe^{\cdots}/Fe^{\cdot\cdot}$ ratios form depends on the activity of H_2S and on pH.

3. As already indicated, the activity of H_2S at low contents of an oxidizing agent is controlled by reaction (37).

These conclusions are entirely sufficient for us to construct the stability fields of iron sulfides.

Stability Diagrams of Sulfides

Unfortunately the values of free energy of amorphous sulfides separating from cold solutions are not available to us. There are data and methods, however, that permit us to calculate these data for hypothetical sulfides and, for the plotted diagrams, at least to evaluate the validity of their existence.

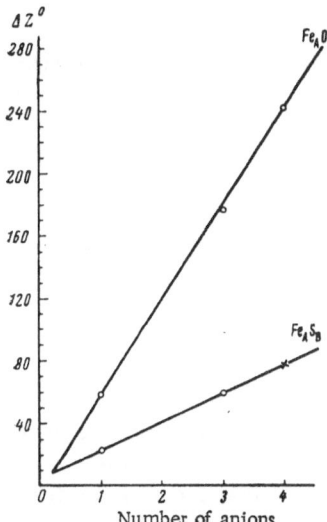

Fig. 37. Dependence of standard free energies of iron sulfides and oxides on the number of sulfur and oxygen atoms.

On the basis of the investigations of Berner [1964b], Polushkina and Sidorenko [1963], and Korolev and Kozerenko [1965], discussed in the preceding section, we may assert with assurance that a cubic sulfide with spinel structure exists. The free energy of this sulfide may be determined by comparing two series of similar compounds: FeO, Fe_2O_3, Fe_3O_4, and FeS, Fe_2S_3, and Fe_3F_4. If for one of these series we plot the free energy along the ordinate axis, and along the abscissa axis we plot some quantitative characteristic in order that, in the transition from one compound to the other, ΔG^0 changes linearly, the change in ΔG^0 for the other series then proves to be linear also [Karapet'yants, 1965].

In other words, if we select such a functional scale that leads to rectification of one dependence, it also leads to rectification of the other dependence. By using this relationship we find the value of $\Delta G^0_{Fe_3S_4}$ by arranging the values of ΔG^0_{FeS} and $\Delta G^0_{Fe_2S_3}$, taken from the handbook of Latimer [1954]. Figure 37 shows one method of calculation. The abscissa axis represents the number of oxygen or sulfur atoms necessary for the given compound. If we accept $\Delta G^0_{FeS} = -23.32$ and $\Delta G^0_{Fe_2S_3} = -59$ kcal as valid, then we obtain $\Delta G^0 = -78$ kcal for Fe_3S_4. As for the other sulfide forms, such as Fe_4S_5, Fe_5S_6, and so on, there appears to be less basis for computing their ΔG^0 values.

For computing the free energies of these sulfides it is necessary to start with a series of more or less plausible assumptions. It may be assumed that any syngenetic sulfide is converted to pyrite during diagenesis, combining elemental sulfur. If, with this, we assume that the free energy of the reaction remains invariant, regardless of the nature of the sulfide, then, on the basis of the reaction

$$\frac{(1-x)}{(1-2x)} FeS_2 = \frac{1}{(1-2x)} Fe_{1-x}S + S; \quad (\Delta G^0 = const)$$

we may write

$$\Delta G^0_{Fe_{1-x}S} = -(2\Delta G^0_{reaction} + \Delta G^0_{FeS_2})\, x + (\Delta G^0_{reaction} + \Delta G^0_{FeS_2})$$

We have thus arrived at a linear equation permitting us to calculate the value of one parameter when another is given. This is the widely known method of comparing two properties under identical conditions in a series of similar substances [Karapet'yants, 1965]: $G' = AG'' + B$. In this case G' is ΔG^0 of the sulfide, and G'' is the characteristic of this sulfide relative to the form of iron. For the case $x = 0$, $\Delta G^0_{reaction} = 16.52$. Whence $\Delta G^0_{Fe_{1-x}S} = 6.80\, x - 23.32$. Substituting the appropriate values of x, we find: $\Delta G^0_{Fe_2S_3} = -63.16$, $\Delta G^0_{Fe_3S_4} = -86.48$. If we use $\Delta G^0_{Fe_2S_3} = -63.16$ kcal in the first method of calculation, then $\Delta G_{Fe_3S_4} = -84.94$, which is 15 kcal lower than the value obtained by the second method of calculation. Thus, the assumption that the free energy of the reaction between pyrite and $Fe_{1-x}S$ remains constant and does not depend on x is confirmed.

In systematizing the great amount of experimental data on the pressure of sulfur above a mixture of pyrite and pyrrhotite, Barnes and Kullerud [1961] explained that the direction of the pyrite ⇌ pyrrhotite reaction is determined completely by the partial pressure of sulfur. They calculated the free energy of this reaction. The thermodynamically probable process was determined by these investigators in the following way:

$$(1+x) FeS_2 = (1-2x) Fe_{1-x}S + {}^1/_2 S_2 \text{ (gas)}$$
$$\Delta G^0 = 26.5 \pm 2 \text{ kcal}$$

Whence, it is easy to calculate

$${}^1/_2 P_{S_2(gas)} = -\frac{\Delta G^0}{1.364} = -\lg K$$

Careful analysis of the equation shows something extraordinary in its form. Actually, in comparing the coefficients for iron, one might conclude that the stoichiometry is preserved only when $x = 0$. If we calculate for sulfur, then, when compensating for the number of atoms, we observe no compensation in the participating electrons. It is shown to be a fact that, in this reaction, apart from gaseous sulfur, which determines the direction of the reaction, gaseous iron also takes part. In experiments on the stability of pyrite in the Fe–S system [Kullerud and Yoder, 1959], it was shown that under equilibrium conditions the pressure above a mixture of pyrite, pyrrhotite, and sulfur is less than above pure sulfur under the same conditions, a fact explained by the partial pressure of iron. Analysis of the liquid phase has shown an insignificant content of iron. The reaction must therefore be written thus:

$$(1+x)\, FeS_2 = (1+2x)\, Fe_{1-x}S + {}^{1}/_{2}S_2\text{(gas)} + 2x^2 Fe\text{(gas)}$$

Then

$${}^{1}/_{2}\lg P_{S_2} = -\frac{\Delta G^0}{1.364} - 2x^2 \lg P_{Fe}$$

However, the effect of the value of $2x^2 \log P_{Fe}$ on the calculation is so slight that it may be neglected. Using the value of ΔG^0_{FeS}, from the equation of Barnes and Kullerud, we find

$$\Delta G^0_{Fe_{1-x}S} = -\frac{23.32 + 39.84x}{(1+2x)}$$

For x = 1/3, ΔG^0 = 21.96, which corresponds to $\Delta G^0_{Fe_2S_3}$ = –65.88 kcal. $\Delta G^0_{Fe_3S_4}$ = –88.74 kcal. If we calculate by the first method, having taken the appropriate value of $\Delta G^0_{Fe_2S_3}$, then $\Delta G^0_{Fe_3S_4}$ = –89.19 kcal, which differs from direct calculation by 0.45 kcal.

Thus, in determining the free energy of formation of cubic sulfide, we may state that the value lies somewhere in the 80–89 kcal range. The difference of 9 kcal naturally has no perceptible effect in constructing the diagrams. A fundamentally valid diagram remains. In our computations below, we shall use the value of $\Delta G^0_{Fe_{1-x}S}$, obtained on the basis of the data of Barnes and Kullerud.

On the basis of the fact that in sedimentary rocks essentially all forms of sulfides may be present, from ferrous sulfides to ferric sulfides, as appears from the previous discussions, we will conclude that x may lie between 0 and 1/3. During the formation of sulfides, one of the following reactions may take place:

$${}^{2}/_{3}Fe(OH)_3 + H_2S = Fe_{2/3}S + 2H_2O + {}^{0}/_{3}S$$

$${}^{3}/_{4}Fe(OH)_3 + {}^{9}/_{8}H_2S = Fe_{3/4}S + {}^{9}/_{4}H_2O + {}^{1}/_{8}S$$

$${}^{4}/_{5}Fe(OH)_3 + {}^{6}/_{5}H_2S = Fe_{4/5}S + {}^{12}/_{5}H_2O + {}^{1}/_{5}S$$

$${}^{5}/_{6}Fe(OH)_3 + {}^{15}/_{12}H_2S = Fe_{5/6}S + {}^{15}/_{6}H_2O + {}^{3}/_{12}S$$

. . . .

In general form, this group of reactions will appear as follows for acidic conditions:

$$(1-x)\, Fe(OH)_3 + 1.5\,(1-x)\, H_2S = Fe_{1-x}S +$$

$$3\,(1-x)H_2O + 0.5\,(1-3x)\, S;$$

and for alkaline conditions

$$(1-x)\, Fe(OH)_3 + 1.5\,(1-x)\, H^{\cdot} + 1.5\,(1-x)\, HS' =$$

$$= Fe_{1-x}S + 3\,(1-x)\, H_2O + 0.5\,(1-3x)\, S.$$

Correspondingly, for acidic conditions:

$$-\lg K = 6.74\,(1-x) - \frac{17.10 + 29.21x}{1+2x} \qquad (45)$$

$$\sum a_S = 10^{4.49 - \frac{11.40 + 19.47x}{(1+2x)(1-x)}}$$

and for alkaline conditions:

$$\sum a_S = 10^{pH - 2.51 - \frac{11.40 + 19.47x}{(1+2x)(1-x)}} \qquad (46)$$

Analysis of the equations we have obtained makes it possible to evaluate the view that the nature of the sulfide that forms from ferric hydroxides depends on Σa_S and pH. In other words, x = $f(H_2S, pH)$. Thus, as a result of thermodynamic computations, we have come to the same conclusions we made from an analysis of Kumai's experiments. The dependence x = $\varphi(H_2S)$ for acidic conditions is shown in Fig. 38. The convexity of the curve attests to the fact that with increase in a_{H_2S} the effect of this factor on the nature of $Fe_{1-x}S$ sulfide becomes progressively greater. That is, the more nearly x → 0, the greater the character of the sulfide depends on variation in a_{H_2S}. By using Eqs. (45) and (46) we may plot graphs on which the relations among three values will be reflected: Σa_S, pH, and x (Fig. 39). In Fig. 39 we have plotted the experimental results of Kumai and have shown the general direction of the process during the course of the experiments.

Naturally we cannot expect precise coincidence of results because we were able to calculate from Kumai's data the residual Σa_S but not the reactive, and the total content of oxidized and reduced forms of iron regardless of whether they are found in the

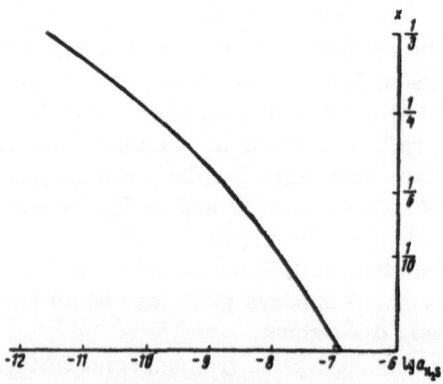

Fig. 38. Probable composition of pyrrhotite of $Fe_{1-x}S$ type in dependence on a_{H_2S}.

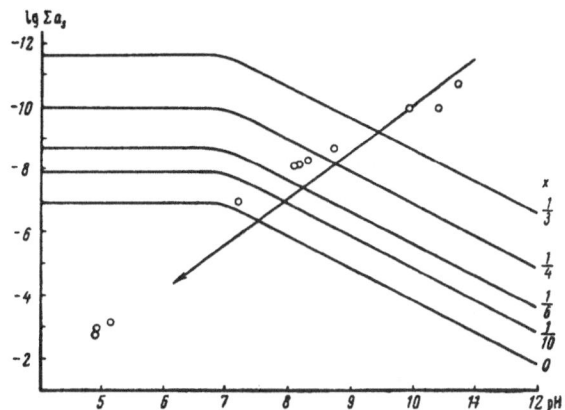

Fig. 39. Explanation of the trend in the processes of Kumai's experiments.

precipitate or in the solution, whereas Fig. 39 reflects the nature of the sulfides in the precipitate with unlimited solubility of Fe¨ under acidic conditions. Nevertheless, despite the qualitative character, Fig. 39 may serve as a good illustration of the

results of Kumai's experiments, showing that with increase of hydrogen sulfide and diminution of pH relative to precipitate of $Fe(OH)_3$, the formation of $Fe_{1-x}S$ sulfide shifts toward $x \to 0$, whereas a low Σa_S and high pH facilitate the formation of sulfide closer to $x \to 1/3$.

The investigated conditions in nature are entirely possible. In normally aerated basins, hydrogen-sulfide contamination begins at some depth from the surface of the sediments, as seen from Fig. 40. This provides for a gradual change of the effect of hydrogen sulfide on ferric hydroxide, and thus makes it possible for sulfides to form with a high content of the ferric ion. But, even under the less favorable conditions of hydrogen-sulfide contamination in the Black Sea it is also possible for Fe_3S_4 to form, as attested by the data of Polushkina and Sidorenko [1963].

When the hydrogen-sulfide system proves to be potential-limiting, then, substituting Σa_S from Eqs. (45) and (46) in Eq. (39), we find how the formation of iron sulfide depends on the oxidation—

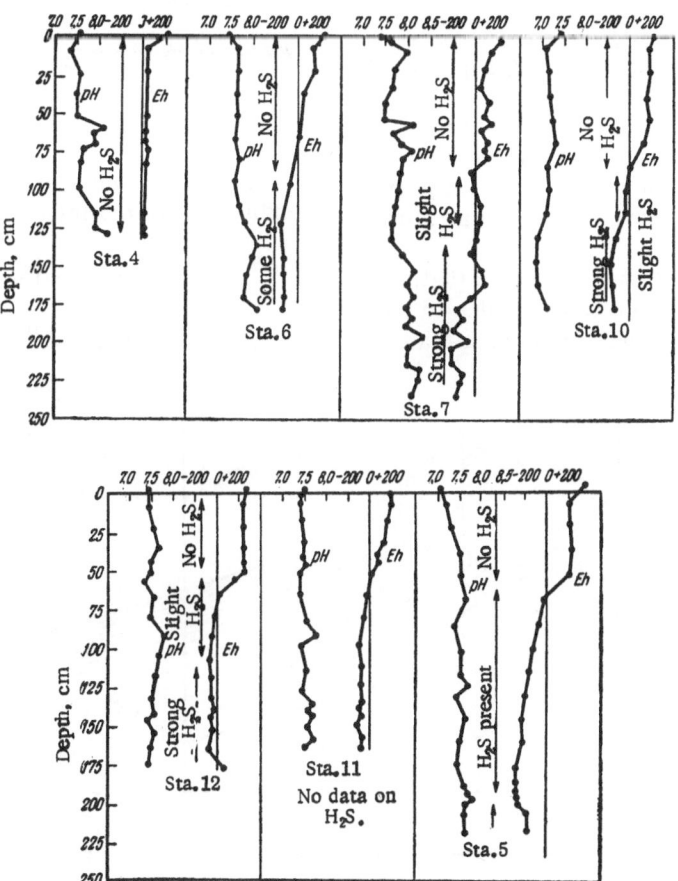

Fig. 40. Relations of pH and Eh in California basin sediments (after Emery and Rittenberg [1952]).

reduction conditions in which the ferric hydroxide occurs:

$$Eh = 0.01 + \frac{0.34 + 0.55x}{(1+2x)(1-x)} - 0.059\,pH \qquad (47)$$

For acidic conditions, when the activity of the ferrous ion may be considerable, it is necessary to use the equation

$$\tfrac{2}{3}Fe^{..} + H_2S = Fe_{2/3}S + 2H^{.} + \tfrac{2}{3}e$$

$$\tfrac{3}{4}Fe^{..} + H_2S = Fe_{3/4}S + 2H^{.} + \tfrac{2}{4}e$$

$$\tfrac{4}{5}Fe^{..} + H_2S = Fe_{4/5}S + 2H^{.} + \tfrac{2}{5}e$$

$$\tfrac{5}{6}Fe^{..} + H_2S = Fe_{5/6}S + 2H^{.} + \tfrac{2}{6}e$$

and so forth.
In the general form:

$$(1-x)\,Fe^{..} + H_2S = Fe_{1-x}S + 2H^{.} + 2xe$$

In similar fashion, after summing the resulting reactions with the reactions of hydrogen–sulfide oxidation to sulfur, we determine

$$Fe_{1-x}S = (1-x)\,Fe^{..} + S + 2\,(1-x)\,e$$

Making simple thermodynamic recalculations, we find

$$Eh = \frac{1 + 1.708x}{(1+2x)(1-x)} \cdot 0.506 - 0.440 + 0.0295\,\lg a_{Fe^{..}} \quad (48)$$

Assigning different values of x within the limits of 0 and 1/3, we may derive a whole series of equations for the formation of sulfides. A particular case, when $a_{Fe^{..}} = 10^{-3}$ mole, is illustrated in Fig. 41.

If any sulfide formed, it would seem that, with decline in oxidation–reduction potential due to the hydrogen-sulfide system, there should be a transi-

tion toward a more reduced form of sulfide. The process may be represented schematically in the following series of equations:

For pH < 7

$$\tfrac{8}{3}Fe_{3/4}S + \tfrac{1}{3}H_2S = 3Fe_{2/3}S + \tfrac{2}{3}H^{.} + \tfrac{2}{3}e$$

$$\tfrac{15}{4}Fe_{4/5}S + \tfrac{1}{4}H_2S = 4Fe_{3/4} + \tfrac{2}{4}H^{.} + \tfrac{2}{4}e$$

$$\tfrac{24}{5}Fe_{5/6} + \tfrac{1}{5}H_2S = 5Fe_{4/5}S + \tfrac{2}{5}H^{.} + \tfrac{2}{5}e$$

. . . .

or in the general form

$$\left(\tfrac{1}{x} - x\right)Fe_{1/(1+x)}S + xH_2S = \tfrac{1}{x}Fe_{1-x}S + 2xH^{.} + 2xe \;\Bigg\}$$

and, correspondingly for pH > 7 $\qquad\qquad (49)$

$$\left(\tfrac{1}{x} - x\right)Fe_{1/(1+x)}S + xHS^{'} = \tfrac{1}{x}Fe_{1-x}S + xH^{.} + 2xe \;\Bigg\}$$

For computing the free energy of the compound $Fe_{1/(1+x)}S$, **x** must be replaced by $1/(1+x)$; then

$$\Delta G^0_{Fe_{1/(1+x)}S} = -23.32 + \frac{6.8}{\frac{1}{x}+3}$$

Whence

$$pH < 7.\ \Delta G^0_{reaction} = \frac{6.80}{(1+2x)(1+3x)} - 16.78\,\frac{(1+2.59x)}{(1+3x)}$$

$$pH > 7.\ \Delta G^0_{reaction} = \frac{6.80}{(1+2x)(1+3x)} - 26.33\,\frac{1+2.74x}{(1+3x)}$$

The oxidation–reduction equations will correspondingly have the form

$$pH < 7,\ Eh = \frac{0.147}{(1+2x)(1+3x)} - 0.364\,\frac{(1+2.59x)}{(1+3x)} -$$

$$-\,0.059\,pH - 0.029\,\lg \sum a_s$$

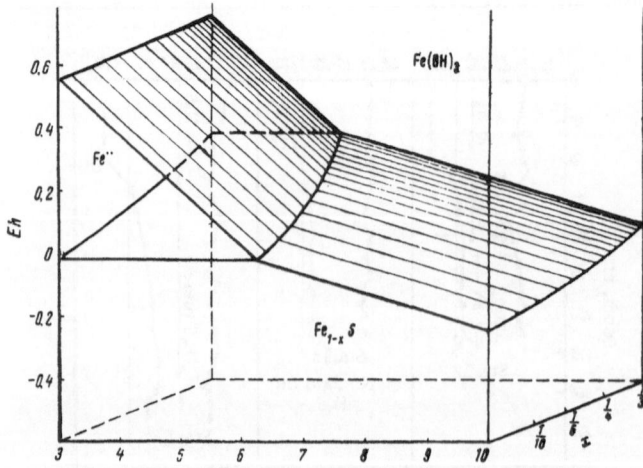

Fig. 41. Stability region of iron sulfides in Eh–pH coordinates.

$$pH > 7, \quad Eh = \frac{0.147}{(1+2x)(1+3x)} - 0.571\frac{(1+2.74x)}{(1+3x)} -$$

$$- 0.029pH - 0.029 \lg \sum a_S$$

If the hydrogen-sulfide system is potential limiting, it is then necessary to use Eq. (39). After substituting the appropriate values of H_2S or HS' from Eq. (39), we find

$$x^2 + 0.930x + 0.072 = 0; \quad x_1 = -0.845, \quad x_2 = -0.085$$

Since, for the investigated conditions, when $x < 0$ is excluded, the resulting solution shows that the transition in trend from Fe_2S_3 to FeS with decline in oxidation–reduction potential of the potential-limiting hydrogen-sulfide system is excluded.

As seen from Figs. 39 and 41, the transition of sulfide from $x = 1/3$ toward $x = 0$ is possible when the decline in pH proceeds farther than may be possible by the hydrogen-sulfide system; or the decline in Eh at the expense of hydrogen sulfide takes place in a buffered system at a single pH value.

The cited analysis makes it possible to understand how the formation of any particular sulfide is determined by the oxidation–reduction potential of the system in which ferric hydroxide occurs and by the pH value. These values are apparently unique parameters, determining the value of x under sterilized laboratory conditions. On the basis of thermodynamic data only, it might be possible to determine the oxidation–reduction conditions for the formation of sedimentary rocks according to the nature of the sulfide found in them. It might even be possible to neglect the activity of iron bacteria, which may change the $Fe^{\cdots}/Fe^{\cdot\cdot}$ ratio in the interstitial water of the muds, and, consequently, somewhat

change the character of the sulfides. As will be seen in the following chapter, subsequent processes accompanying the consolidation of the sediments introduce such essential modifying factors that the determination of Eh and pH according to kind of sulfides is impossible. We must be satisfied for the time being with the fact that Eqs. (47) and (48) may mathematically describe all actual variants of the formation of $Fe_{1-x}S$ in dependence on x, pH, Eh, and $a_{Fe^{\cdot\cdot}}$.

To what we have said previously we might add the interesting remarks of F. Feigel', who, on the basis of experiments, concluded that, at high concentrations of $Fe^{\cdots}$ and low concentrations of the sulfide ion, FeS forms chiefly, and at low concentrations of $Fe^{\cdots}$ and high concentrations of the sulfide ion, Fe_2S_3 forms. Since the solubility of $Fe(OH)_3$ declines appreciably with increase in alkalinity, as is clear from Eq. (1), under alkaline conditions Fe_2S_3 will certainly form, and under acidic conditions FeS. This conclusion is in agreement with the conclusion of Latimer [1952], according to which Fe_2S_3 should not oxidize the sulfide ion to sulfur in an alkaline solution. Kumai [1958], on the basis of the cited experiments, expanded the possibilities of variations among sulfides and stated that the formation of FeS or Fe_2S_3 as a result of reaction between ferric hydroxide suspended in the water and hydrogen sulfide is determined to a considerable extent by the rate of formation of Fe_2S_3 precipitate and the rate of $Fe^{\cdots}$ reduction under the influence of hydrogen sulfide. Consequently, the composition of the reaction product may be different according to the pH of the suspension, the amount of hydrogen sulfide entering the reaction, the method of introducing the additives to the reacting mass, and a number of other factors.

CHAPTER 4

CONDITIONS FOR FORMATION OF IRON MINERALS DURING SEDIMENTATION AND DIAGENESIS

In the preceding chapter we examined the problem of constructing diagrams of minerals when the initial material consists of ferric hydroxide. We also analyzed some particular questions of mineral paragenesis. These data have made it possible to gain some understanding of the fact that the formation of any particular mineral is more complex, perhaps taking place in many different ways, than previously thought. It is necessary to find some agreement among these data and to draw a complete picture of the formation and successive changes in authigenic minerals during sedimentation and diagenesis. The material in the third chapter cannot alone furnish all we need. To create the full picture of the formation of authigenic minerals and their alterations during diagenesis, we must turn to the geologic facts and, on the basis of these, use the constructions we derived in Chap. 3.

By diagenesis we mean the transformations of primary components in the sediments under the thermodynamic conditions of the earth's crust. For diagenesis, the nature of the successive shifting of equilibrium conditions is due to compaction, change in gas conditions, change in intensity of bacteriological factors, and still other things. Minerals are altered by virtue of these changes in conditions. In time relations, diagenesis is characterized by progressive attenuation of the processes.

The parallel comparison of thermodynamic data with the geologic facts seems to us the most reliable course, permitting us to make an objective evaluation of the possibility of using the thermodynamic method as well as a consideration of the aspects of mineral formation, in addition to supplying us with criteria that may be used in determining the oxidation–reduction conditions under which rocks form. This composite method permits us to avoid the errors that appear during separate examination of thermodynamic and geologic factors. As may have become clear already, separate examination of these factors into definite stages of generalization takes on a formal character, leading progressively to greater divergence between the conclusions and reality.

Let us recall, for example, the purely geologic approach to interpretation of the relations of elements having different valencies in a hydrochloric-acid extract for determining the oxidation–reduction potential. This approach led to an attempt to set up the analytical results in the Nernst equation, causing a contradiction between computed and actually measured oxidation–reduction potential. The purely physicochemical approach is represented by the diagram of Garrels. This approach does not take into account the process actually taking place in nature in the formation and oxidation of hydrogen sulfide. Therefore, despite formally proper methods of computation, the resulting diagrams do not reflect even present-day sedimentation. In constructing the diagrams, we tried to come to the form

$$Eh = f(pH, m, n \ldots)$$

In particular, the value of Eh appears as a variable dependent on several parameters for the calculation of any particular mineral that may form from ferric hydroxides. However, whereas it is not known what the mineralogical state may be in sediments during their formation and consolidation, i.e., whereas we may not know what surface of $Eh = f(pH, m, n \ldots)$ it is necessary to select, the value of Eh becomes an independent variable, and the formation of any particular mineral becomes a function of Eh, pH, m, n . . . . Therefore, before beginning to construct mineral diagrams reflecting natural processes it is advisable to examine the behavior of all the least necessary numbers of independent variables (including Eh) that participate in the formation of minerals. Below we shall consider basins in the water and mud of which sulfate ion is being reduced. Basins in which the sulfate ion is not being reduced are a special case.

Change of pH during the Diagenesis of Sediments

The pH value changes comparatively little during diagenesis. According to Strakhov [1960b], the pH in marine carbonate sediments always remains at the 7.2−7.8 level, apparently because of buffering in the system. For the Sea of Okhotsk and Bering Sea, some increase in pH with depth in the layer of mud is observed [Bruevich, 1956; Bruevich and Zaitseva, 1958]. In sediments of the Gulf of California, pH undergoes small fluctuations, both upward and downward relative to that of the surface mud [Zobell, 1946]. In the oceanic sediments of the northeastern part of the Pacific Ocean, the pH remains approximately constant [Romankevich and Petrov, 1961]. Variations of pH in different directions have been reported from the sediments of the Indian Ocean according to depth of burial in the muds (pH values within the limits of 6.91 and 7.69 according to Zheleznova and Shishkina [1964]). The pH is commonly near 7 in the muds of fresh-water lakes.

Change of Oxidation − Reduction Potential during Diagenesis of Sediments

The concept of an oxidation−reduction boundary is closely related to the concept of mineral indicators. The essence of it is that in a basin where sediments are accumulating there must exist an oxidation−reduction boundary at which there is equilibrium of reducing and oxidizing processes. Downward we should observe reducing conditions. Upward the environment becomes increasingly oxidizing. Logically, this concept is fully convincing, since an oxygen zone exists above the basin and microbiological processes are at work in the muds, strongly suppressing oxidizing agents (or oxygen).

Apart from this, thanks to the activity of microorganisms, the medium is progressively enriched in reducing agents, such as hydrogen, the sulfide ion, the reduced series of hydrocarbons, and others. Extrapolations of the increase in degree of reduction downward in the sediments therefore agree outwardly to be very proper, even if the basis for it is not sufficiently established. Actual data, however, indicate that this view does not accord with the true picture of change in Eh. For example, if we examine the microbiological factors as a principal cause for development of reducing conditions, the eye is immediately struck by the circumstance that microorganisms are active only in the upper layer of muds, in a layer 20-25 cm thick. Below this layer their activity declines sharply and, consequently, the sediments are deprived of the prin-

cipal agent that might increase the reducing capability of the environment.

Therefore, in order to be consistent, it is necessary at the very least to assume that the reducing capacity of the environment must, with increasing depth in the sediment, remain at the level of that 20- to 35-cm layer of muds in which microbiological processes are intense. Even with this approach, which is not very complex, extrapolation on the decline of oxidation−reduction potential with depth has proved to be insufficiently reliable.

The deeper we pursue this question and examine the fundamental mineral transformations that take place in the layer of muds of present-day basins, the more we must come to the single conclusion, contradicting the assumed extrapolation: according to mineral transformations that take place in the muds, processes must be at work that lead to increase rather than decrease in oxidation−reduction potential. This is attested, in particular, by the change of the hydrotroilite type of FeS to pyrite, noted by many investigators. It is assumed that pyrite is a more highly oxidized form of FeS than hydrotroilite.

Results from drilling in recent muds in seas, oceans, and lakes, with direct measurements of Eh, are the most direct and reliable facts that permit us to go beyond the realm of hypothesis and supposition. These direct facts confirm the view that the oxidation−reduction potential increases downward from the thin layer of muds in which microbiological processes are most active. A rather large number of such facts have now accumulated.

The increase in oxidation−reduction potential with depth, as compared with the upper thin layer of muds, is indicated by the data of Ovsyannikova [1951]. In studying the distribution of microorganisms in the muds of saline lakes, she did not assign this fact the importance that, in the hands of geochemists, might serve as a key to understanding the authigenic formation of minerals. Ovsyannikova, like a number of other investigators [Ekzertsev, 1948; Kuznetsov, 1952], associated this phenomenon entirely with the decline in activity of microorganisms.

Increase in oxidation−reduction potential has been noted in California basin sediments (according to Emery and Rittenberg [1952]; see Fig. 40). In examining the indicated figure, it is not hard to convince oneself that decline in oxidation−reduction potential takes place in that part of the mud in which hydrogen sulfide appears. Lower, despite the fact that hydrogen sulfide does not disappear, a more or less perceptible increase in oxidation−reduction potential is noted.

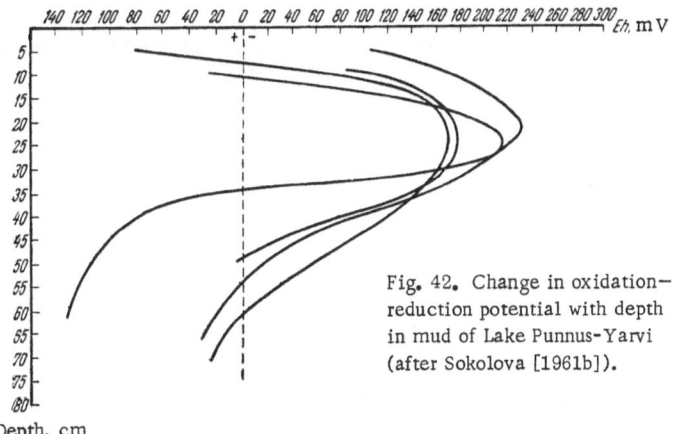

Depth, cm

Fig. 42. Change in oxidation–
reduction potential with depth
in mud of Lake Punnus-Yarvi
(after Sokolova [1961b]).

Strakhov [1954], in analyzing data on recent sediments, came to the conclusion that the pattern of change in Eh through a vertical section of sediment is still not completely clear. With increasing depth in the reducing zone, Eh generally more or less rises again. In an earlier work Strakhov [1953] extrapolated the increase in potential to zero. A well-defined increase in oxidation conditions with depth in mud was noted in the sediments of Lake Punnus-Yarvi (Fig. 42). Debyser [1957] observed an increase in Eh with depth in sediments of the Baltic Sea. Measurements of Eh and pH in the central part of the Pacific Ocean by Rittenberg show that Eh (in a core) generally increases with depth [Baas Becking et al., 1960]. A beautiful confirmation of the increase in oxidation—reduction potential along a vertical section of mud as compared with the upper thin layer is found in data on the Sivash. The pattern is clearly traced here in all boreholes (Fig. 43).

Similar phenomena may be observed also in muds of the Indian Ocean. Gordeev [1962], who studied the distribution of pH and Eh in these muds, has cited the following facts: in the zone of iceberg muds, a column 2.3 m long, represented by uniform clay mud, exhibits a variation in pH between 7.55 and 7.65 and a range in Eh from +17 to +33 mV. In the zone of foraminiferal muds, uniform globigerina ooze (a core 2.7 m long) shows a range in pH from 7.15 to 7.30 and a range in Eh from +148 to +167 mV. Numerous measurements in the zone of perfectly uniform abyssal red clays indicate that down to a depth of 20 cm the pH is 7.25–7.15 and the Eh is +65 to +115 mV. At depths somewhat greater than 2 m below the surface of the mud, pH is 6.6–6.8 and Eh ranges from −52 to −30 mV. Still deeper, pH continues to decline, but Eh increases to +15

mV. In the zone of clayey-foraminiferal ooze, 20 measurements of pH and Eh were made on a core 13 m long. The upper 20 cm consisted of brownish red clay, changing to bluish gray clay below. The upper layer had a pH of 7 and an Eh of +205 mV. Below this 20-cm layer the pH was observed to increase to 7.6–7.8 and the Eh to +440 (+470) mV. On the basis of such data, the author of the paper concluded that some increase in the oxidation—reduction potential to a value near zero indicates that both these processes (meaning oxidation and reduc-

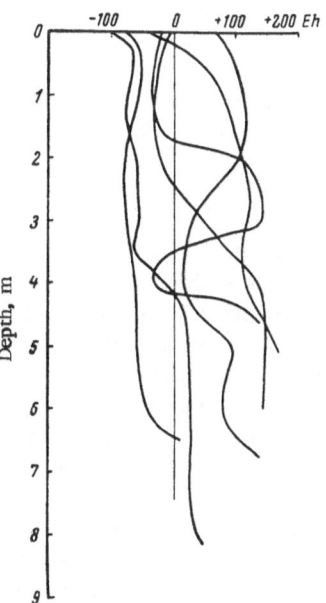

Fig. 43. Change in oxidation–reduction potential with depth in muds of
the Sivash (after Stashchuk, Suprychev,
and Khitraya [1964]).

tion) are in a state near equilibrium, a conclusion that is erroneous, as we pointed out in the second chapter.

According to Zheleznova and Shishkina [1964], a more or less perceptible increase in oxidation-reduction potential begins downward from the thin layer of most active reduction in muds of the Indian Ocean. And, although the authors of the previous paper make no mention of such a phenomenon, it stands out rather clearly, as seen in their Table 1. One's attention is drawn to the fact that the increase in oxidation-reduction potential begins immediately after the appearance of hydrotroilite in the sediments. According to Shishkina [1961] a similar picture has been observed in a vertical section of muds in the Black Sea: an increase in oxidation-reduction potential with increase depth in the mud. Shishkina explained this behavior of the potential by the glacial period.

It seems to me that, against the background of all the cited examples, this phenomenon must be considered an element in the general tendency characteristic of all recent sediments. There is apparently some powerful factor that leads to a change in the direction of the oxidation-reduction processes during the period of diagenesis. This factor is the cause of a general systematic increase in oxidation-reduction potential, taking place simultaneously with the diagenetic trend.

It is characteristic that the phenomenon described above is manifested most clearly in muds containing hydrogen sulfide, under conditions, it would seem, where there is no cause for oxidizing processes. This unexpected factor places in doubt all predictions of mineral transformations based on the assumption of continuous increase in reducing capacity of muds during diagenesis. It is for this reason that many facts attesting to increase in oxidation-reduction potential with depth are explained by some local cause without note being taken that the number of exceptions is considerably greater than the number of cases fitting the expected pattern based on unfounded extrapolation.

If we do not base our expectations on this extrapolation but compare all exceptions observed in muds of different kinds and different ages occurring in completely different basins with no geographic links, we cannot arrive at any general cause responsible for the distinctive tendency toward increase in oxidation-reduction potential with progress of diagenesis. This cause is hidden in the specific aspects of diagenesis itself: the conversion of more soluble compounds to less soluble.

At present we have not yet sorted out all the complex knowledge we have concerning oxidation-reduction systems in sediments, and it is therefore difficult to define the process from all points of view. For now we must be satisfied with the awareness that the view of increase in oxidation-reduction potential during the course of diagenesis is based on numerous and rather clear facts.

The Distribution of Hydrogen Sulfide

There are comparatively few data concerning the distribution of free hydrogen with depth in muds, and, unfortunately, even these data cannot be fully used since there is, as yet, no thoroughly developed method of reliable determination of this gas in the interstitial water of muds. Methods based on the displacement of hydrogen sulfide directly from the mud by inert gas [Volkov, 1959; Kaplan, 1963] or on squeezing the hydrogen sulfide from the muds (centrifuging) cannot be considered reliable, as Berner [1964a] validly pointed out, if only because sulfur is represented chiefly by the HS' ion at the pH values of marine muds (see Fig. 34). Therefore, for complete and reliable extraction of H_2S from a medium, it is necessary to acidify it somewhat, but this causes decomposition of sulfides such as pyrrhotite, and this leads to values of H_2S content that are too high. These circumstances must be considered in determining the distribution of hydrogen sulfide, and we must not try to use the distribution as a quantitative index.

In studying the distribution of the forms of sulfur in muds of the Gulf of California, Berner [1964a] first called notice to the fact that the muds contain no sulfides but pyrite. Therefore, as Berner noted, the difficulties that arise when determining different forms of sulfur were bypassed because of accidental circumstances. Only one core was characterized by active reduction of sulfate (L 154) and, consequently, by the appearance of hydrotroilite at a depth of 50 cm below the surface of the sediments. It was this core that was distinguished by the highest content of hydrogen sulfide at the surface of the mud and was characterized by gradual diminution of hydrogen sulfide vertically with simultaneous increase in Eh. According to Berner, this distribution must reflect the fundamental behavior of hydrogen sulfide in muds in general, but if diffusion along the floor is sufficiently strong the distributional pattern of hydrogen sulfide is destroyed.

A decrease in hydrogen-sulfide content with depth in the sediments is observed in muds of the

Black Sea. A number of investigators [Strakhov, 1959; Shishkina, 1962; Volkov, 1964] believe that many changes taking place in muds of the Black Sea, as one goes from younger to older muds, are related to climatic changes, particularly to effects of the glacial period. We shall not discuss this question here, since it represents fundamentally but one of the variants for explaining the data uncovered by investigators. But, keeping in mind the fact that this point of view exists, let us consider not the absolute change in hydrogen-sulfide content but the relative change in reference to total reduced sulfur. The nature of the change is shown in Fig. 44, which was constructed from data by Volkov [1961a]. The figure leads us to conclude that the content of hydrogen sulfide declines downward in the sediment.

It is simplest to associate this process with two phenomena: decrease in activity of sulfate-reducing microorganisms and the occurrence of hydrogen-sulfide diffusion. As experiments of Ivanov and Terebkova [1959] and Sokolova and Sorokin [1958] with tracer atoms of sulfur have shown, downward from the surface of the active layer in which sulfate reduction is most intense there occurs a rapid decline in rate of H_2S formation, although the number of sulfate-reducing bacteria does not diminish.

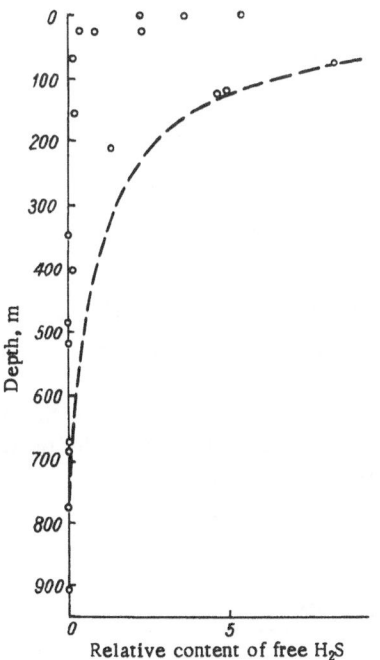

Fig. 44. Relative decline in content of free hydrogen sulfide with depth in muds of the Black Sea (from data of Volkov [1961a]).

Analysis of Mineral Formation

If we compare the two investigated phenomena (the increase in oxidation–reduction potential downward in the sediments and the parallel decline in content of hydrogen sulfide), the conclusion that these processes must be interconnected somehow is forced on us. The pattern of this interconnection may be best determined by comparing directly the Eh and the hydrogen-sulfide content in muds. However, because of the fact that at present there are no sufficiently reliable methods of determining hydrogen sulfide in the presence of sulfides readily decomposed by acid, this comparison proves to be ineffective. The nature of the pattern must be therefore explained by means of indirect data.

The simplest explanation of the fact that in marine sediments the hydrogen-sulfide content declines along with rise in oxidation–reduction potential downward in the sediments may be given by assuming that the hydrogen-sulfide–sulfur system determines the potential. Actually, if we analyze Eq. (39) we may convince ourselves that a decline in a_{H_2S} leads to an increase in oxidation–reduction potential. As Skopintsev [1957] has shown, it is this system that determines the oxidation–reduction potential in the hydrogen-sulfide zone in the Black Sea.

Experiments of E. V. Rozhkova, É. G. Kuznetsova, and É. G. Vasil'eva on the appearance of sulfate-reducing bacteria in sedimentary rocks with parallel measurement of Eh fully confirm the fact that Eh is regulated by the H_2S content. We may therefore justly state that increase in Eh, taking place along with decrease in a_{H_2S} in recent marine sediments, may be related to the fact that the hydrogen-sulfide– sulfur system determines potential.

The proposed hypothesis acquires force from the aspect of most active reduction of sulfates. In the Black Sea, at depths below 150–200 m, the zone in which the hydrogen-sulfide system proves to be a potential-determining system begins at the surface of the mud. Where admission of oxygen to the mud is sufficient, this zone begins at a certain depth within the mud, as observed, for example, in sediments of California basins (see Fig. 40). This last variant describes the general case, whereas the muds in the deep part of the Black Sea reflect a particular example. From the position of the general case, let us examine the specific aspects of the transitional zone lying above the zone where the hydrogen-sulfide system is the potential-determining system.

As Strakhov has noted [1956, 1960b], there are generally two physicochemical environments in any basin, distinguished one from the other. One possesses a greater or lesser reserve of oxygen, and below it occurs an environment with a high content of hydrogen-sulfide. The boundary between the two must be drawn by the drop in potential characterizing the transition to the hydrogen-sulfide potential-determining system. This division, which appears correct in principle, proves to be somewhat over-simplified on closer examination. The fact is that the existence of a boundary marked by a potential difference by no means indicates a complete absence of hydrogen sulfide above this boundary or the complete absence of oxygen below it. We have facts that attest to the existence of hydrogen sulfide in a definite zone above the transitional boundary in amounts greater than that permitted by Eq. (39). The experiments of Levchenko, analyzed by us in Chap. 3, in which the effect of oxygen on the oxidation of hydrogen sulfide was analyzed (see Fig. 35), confirm this relationship. Support is also found in the very fact of sulfur-bacteria activity, most of which leads to excess hydrogen sulfide and a non-equilibrium oxidation—reduction potential in system (39).

A beautiful example of the analyzed phenomenon is found in muds of the Sivash. This basis is characterized by depths of about 2 m. Intense mixing, the surge and swash of the water, is responsible for saturating the surface part of the mud with oxygen. On the other hand, intense reduction of sulfates restores the reserve of consumed hydrogen sulfide. This leads to nonequilibrium content of hydrogen sulfide, as a consequence of which hydrotroilite is formed at high Eh values.

By using the cited material and the conclusions obtained in analysis of the oxidation—reduction state of the hydrogen-sulfide system (see Chap. 3), we may state that in a basin in the muds (and water) of which sulfate-reducing processes are at work the following specific zones exist:

1. A zone where the potential-determining system is a strong oxidizing system and oxygen is present;

2. A transitional zone in which the potential-determining system is a strong oxidizing system but in which there occurs an increase, as compared with Eq. (39) in the nonequilibrium content of hydrogen sulfide;

3. A zone in which the potential-determining system is the hydrogen-sulfide system; and

4. A zone below the three indicated above, in which the hydrogen-sulfide system again loses its role as the potential-determining system.

Material arriving from land passes through all these zones sequentially, undergoing corresponding transformations. Depending on the specific character of a basin and its sediments, each of these zones may expand or contract.

Let us consider from this viewpoint the formation of minerals according to oxidation—reduction potential, pH value, and nature of the zone. We shall begin with the simplest case, when ferric hydroxide and hydrogen sulfide are present, but carbon dioxide is absent. In this case we may suppose that ferric oxide, magnetite, and sulfides exist. Let us consider the possible paragenesis of magnetite and iron sulfide. The particular case of sulfides forming from ferric hydroxide, when hydrogen sulfide is present as the only reducing agent, was examined in the analysis of Kumai's experiments. The general case, when the reducing agent is unknown, may be expressed by the following reactions:

$$^2/_3 Fe(OH)_3 + H_2S + {}^0/_3 H^{\cdot} + {}^0/_3 e = Fe_{2/3}S + {}^6/_3 H_2O$$

$$^3/_4 Fe(OH)_3 + H_2S + {}^1/_4 H^{\cdot} + {}^1/_4 e = Fe_{3/4}S + {}^9/_4 H_2O$$

$$^4/_5 Fe(OH)_3 + H_2S + {}^2/_5 H^{\cdot} + {}^2/_5 e = Fe_{4/5}S + {}^{12}/_5 H_2O$$

$$^5/_6 Fe(OH)_3 + H_2S + {}^3/_6 H^{\cdot} + {}^3/_6 e = Fe_{5/6}S + {}^{15}/_6 H_2O$$

. . . .

Whence we obtain the general form of the reaction

$$(1-x)\,Fe(OH)_3 + H_2S + (1-3x)\,H^{\cdot} + (1-3x)\,e =$$
$$Fe_{1-x}S + 3(1-x)\,H_2O.$$

By combining this equation with the one reflecting the conditions of magnetite formation from ferric hydroxide (9), we obtain

$$3Fe(OH)_3 + \frac{3}{1-x} H_2S + \frac{3(1-3x)}{1-x} H^{\cdot} + \frac{3(1-3x)}{1-x} e = \frac{3}{1-x} Fe_{1-x}S + 9H_2O$$

+

$$Fe_3O_4 + 5H_2O = 3Fe(OH)_3 + H^{\cdot} + e$$

$$Fe_3O_4 + \frac{3}{1-x} H_2S + \frac{2(1-4x)}{1-x} H^{\cdot} + \frac{2(1-4x)}{1-x} e = \frac{3}{1-x} Fe_{1-x}S + 4H_2O. \qquad (50)$$

The resulting reaction (50) defines the conditions for the change of magnetite to iron sulfide according to the oxidation—reduction of the environment, the pH value, and a_{H_2S}. If the system of hydrogen sulfide determines the potential, then reaction (50) must lead to agreement with reaction (38):

$$\frac{3}{1-x}\,\text{Fe}_{1-x}\text{S}+4\text{H}_2\text{O}=\text{Fe}_3\text{O}_4+\frac{3}{1-x}\,\text{H}_2\text{S}+2\,\frac{1-4x}{1-x}\,\text{H}^{\cdot}+2\,\frac{1-4x}{1-x}\,e$$

$$+\qquad \frac{3}{1-x}\,\text{H}_2\text{S}=\frac{3}{1-x}\,\text{S}+\frac{6}{1-x}\,\text{H}^{\cdot}+\frac{6}{1-x}\,e$$

$$\overline{\qquad\frac{3}{1-x}\,\text{Fe}_{1-x}\text{S}+4\text{H}_2\text{O}=\text{Fe}_3\text{O}_4+\frac{3}{1-x}\,\text{S}+8\text{H}^{\cdot}+8e.\qquad}$$

The ordinary method of calculation leads to the following equation for the oxidation−reduction potential for the paragenesis of sulfide and magnetite:

$$Eh=0.38\,\frac{1+1.71x}{(1+2x)\,(1-x)}-0.03-0.059\,pH \qquad (51)$$

As seen from the resulting reaction, the equation specifies magnetite as an oxidized form relative to all varieties of sulfides. Therefore, the equilibrium plane reflecting Eq. (51), at all values of x, must lie below the plane of magnetite formation from ferric hydroxide. Let us compare these planes. The equation of ferric hydroxide−magnetite (18) was derived in the preceding chapter. For this paragenesis, Eh = 0.31−0.059pH. Equation (51) for x = 0 is transformed to the equation Eh = 0.35 − 0.059pH, and for x = 1/3 it is described by the condition Eh = 0.50 − 0.059pH. Intermediate values of x will be included, correspondingly, within the limits 0.35 < $< E^0 <$ 0.50.

As this analysis shows, for all values of x the magnetite−sulfide paragenetic plane lies above the magnetite−ferric hydroxide plane. From this we may conclude that, when the potential-determining system is the hydrogen-sulfide−sulfur system, magnetite does not form as an authigenic mineral. Under marine conditions, i.e., where sulfate ions are present in the water and sulfate-reducing bacteria are active, the first mineral that forms from ferric hydroxide is iron sulfide. According to what environment relative to $a_{\text{H}_2\text{S}}$ (and, consequently, to Eh) the ferric hydroxide falls into, a variety of Fe_{1-x}S sulfides will form. Only siderite may form along with the iron sulfides. However, the possibilities for the formation of this mineral are extremely limited, as the following analysis will attest.

The condition for equilibrium of siderite with sulfides may be expressed by the reactions

$$^2/_3\text{FeCO}_3+\text{H}_2\text{S}=\text{Fe}_{^2/_3}\text{S}+^2/_3\text{CO}_2+^2/_3\text{H}_2\text{O}+^2/_3\text{H}^{\cdot}+^2/_3e$$

$$^3/_4\text{FeCO}_3+\text{H}_2\text{S}=\text{Fe}_{^3/_4}\text{S}+^3/_4\text{CO}_2+^3/_4\text{H}_2\text{O}+^2/_4\text{H}^{\cdot}+^2/_4e$$

$$^4/_5\text{FeCO}_3+\text{H}_2\text{S}=\text{Fe}_{^4/_5}\text{S}+^4/_5\text{CO}_2+^4/_5\text{H}_2\text{O}+^2/_5\text{H}^{\cdot}+^2/_5e$$

. . . .

In general form, the reaction is written in the following form:

$$(1-x)\,\text{FeCO}_3+\text{H}_2\text{S}=\text{Fe}_{1-x}\text{S}+(1-x)\,\text{CO}_2+$$
$$(1-x)\,\text{H}_2\text{O}+(2\cdot x)\,\text{H}^{\cdot}+(2x)\,e$$

Since we based this analysis on the assumption that the hydrogen-sulfide−sulfur system was potential determining, after substituting Eq. (39) we obtain

$$\frac{1}{1-x}\,\text{Fe}_{1-x}\text{S}+\text{CO}_2+\text{H}_2\text{O}=\text{FeCO}_3+\frac{1}{1-x}\,\text{S}+2\text{H}^{\cdot}+2e$$

Whence

$$Eh=\frac{0.506+0.864x}{(1+2x)\,(1-x)}-0.218-0.059\,pH-0.0295\lg P_{\text{CO}_2} \qquad (52)$$

Equation (52) makes it possible to compute the conditions for the conversion of siderite to sulfide. However, to use it, it is necessary to verify that the first mineral forming from ferric hydroxide is siderite, since, if the sulfide forms directly from the hydroxide, the above equation loses its significance. In other words, it is necessary to verify that the plane indicating the oxidation−reduction conditions for the formation of siderite will be higher relative to the value of the oxidation−reduction potential than the plane reflecting the conditions for sulfide formation throughout the entire range of acidic−alkaline conditions. The paragenesis of ferric hydroxide and siderite is indicated in Eq. (35), and of siderite and iron sulfides in Eq. (52). A positive difference between these equations indicates the possible existence of the series

$$\text{Fe(OH)}_3 \rightarrow \text{FeCO}_3 \rightarrow \text{Fe}_{1-x}\text{S} \rightarrow \text{FeS}_2$$

A negative difference will indicate that the formation of siderite is theoretically improbable, and the series will be

$$\text{Fe(OH)}_3 \rightarrow \text{Fe}_{1-x}\text{S}$$

The difference will be designated by ΔEh:

$$Eh=0.466-0.059\,pH+0.059\lg P_{\text{CO}_2}$$

$$Eh=\frac{0.506+0.864x}{(1+2x)\,(1-x)}-0.218-0.059\,pH-0.0295\lg P_{\text{CO}_2}$$

$$\overline{\Delta Eh=0.684-\frac{0.506+0.864x}{(1+2x)\,(1-x)}+0.089\lg P_{\text{CO}_2}}$$

The formation of siderite proves to be possible when $\Delta Eh > 0$. When $\Delta Eh \rightarrow 0$ the possibility of siderite forming becomes progressively more limited and, finally, when $\Delta Eh = 0$, it practically disappears. In this limiting case

$$\lg P_{CO_2} = \frac{5.68 + 9.71x}{(1 + 2x)(1 - x)} - 7.68$$

The dependence $\lg P_{CO_2} = f(x)$ is shown in Fig. 45. On the basis of this equation, we may conclude that if in a mineral-forming environment the sulfide ion is present and the hydrogen–sulfide–sulfur system determines potential, then, with Fe_2S_3 forming, siderite generally will not. In all other cases, siderite may be the first mineral, but even under the most favorable conditions its formation is possible only when $P_{CO_2} \geqslant 10^{-2}$ atm. If this condition is not realized, sulfide proves to be the first mineral forming from ferric hydroxide. Since, in going from alkaline to acidic conditions and from oxidizing to reducing conditions, the probability of FeS forming rather than Fe_2S_3 increases, then the probability that siderite will form increases in the same direction. But, regardless of these conditions, primary siderite in marine rocks must indicate that P_{CO_2} was no lower than 10^{-2} atm. On the other hand, the presence of primary sulfides in sediments (including pyrite, as we have discovered) indicates conditions such that $P_{CO_2} < 10^{-2}$ atm.

The specific diagram for the paragenesis of siderite–sulfide–hydroxide at $a_{Fe^{..}} = 10^{-3}$ mole is shown in Fig. 46. For modern seas and oceans in which $P_{CO_2} \approx 10^{-3} - 10^{-4}$ atm [Sillen, 1965; Krumbein and Garrels, 1952; Strakhov, 1960], the formation of primary siderite proves to be impossible

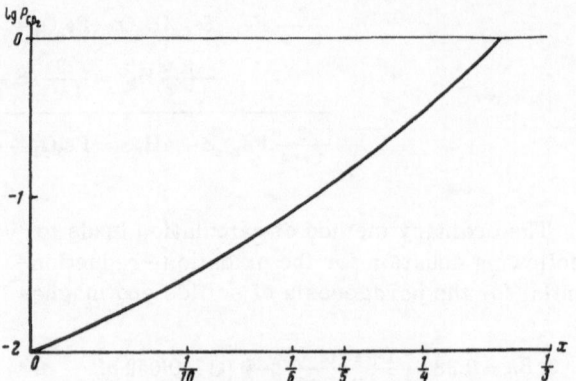

Fig. 45. The value of P_{CO_2} necessary for the paragenesis sulfide–siderite according to different types of sulfides.

regardless of the sulfide that forms. From this we may conclude that during reduction the first mineral that ferric hydroxide is changed into after falling into modern marine sediments proves to be a sulfide.

This is actually confirmed by facts. As is well known, in modern marine sediments siderite has been nowhere detected. References in the literature in regard to this mineral and its relations to recent sediments cannot be considered sufficient basis for belief. For example, Romankevich and Petrov [1961] have written concerning siderite in muds of the northeastern part of the Pacific Ocean, but no mineralogical data are given in this paper. From the data given, it is seen that the actual existence of siderite is maintained on the basis of the sequential precipitation of authigenic minerals that

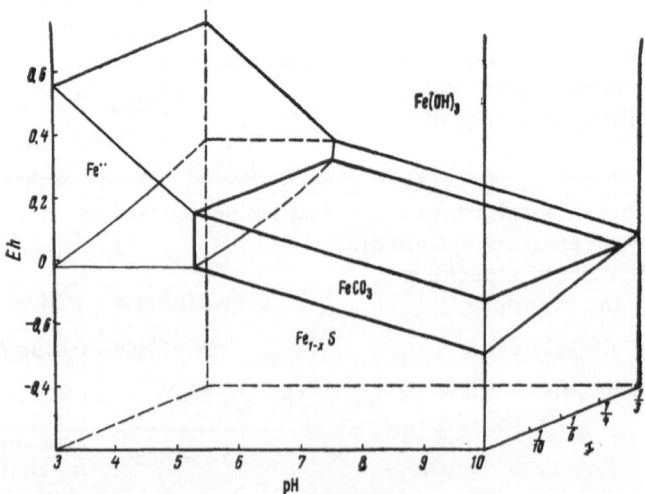

Fig. 46. Stability diagram for ferric hydroxide, siderite, and sulfide at $a_{Fe^{..}} = 10^{-3}$ mole.

should exist if the series of mineral indicators proposed by Strakhov [1953, 1959, 1960] is correct.

Romm [1950] and Strakhov [1954] recalculated divalent iron determined in sediments of basins on the Taman' Peninsula to siderite (iron that could not be bound in sulfide). The basis for this was the diagram of Krumbein and Garrels. And this gives us the right to say that in this case the problem of the existence of siderite is not yet clearly solved. Nevertheless, Skopintsev [1957], basing his conclusion on the work of Strakhov, considered the existence of siderite an obvious fact. Apart from the above discussions, there still are reports of the formation of siderite in recent sediments, but they are all based on similar indirect (and not sufficiently supported) proofs. Study of the forms of iron in muds of the Black Sea, accompanied by determination of CO_2 and $Ca^{..}$, permitted Strakhov [1959] to draw the following eye-opening conclusion: "as for chemical analysis, in all samples so analyzed, it was impossible to find any carbonate of iron (or magnesium). CO_2 scarcely sufficed for binding merely the CaO. And these relations obtain not only in sediments of the oxygen zone but in deposits of the hydrogen-sulfide zone as well."

As far as we know no direct discoveries have been made of siderite in recent sediments. Putting this together with the fact that in modern basins $P_{CO_2} < 10^{-2}$ atm, the absence of siderite should cause no amazement. The absence of siderite and magnetite and the formation of iron sulfides directly from hydroxides, as observed in all basins in which the water contains sulfate ions, should serve as still one more firm proof that the hydrogen-sulfide-sulfur system determines potential in marine deposits.

Let us trace the further fate of sulfides that have formed in the sediments. In the presence of elemental sulfur it is thermodynamically possible to convert any sulfide to pyrite. This change is most likely to proceed along the course illustrated by the following reactions:

$$3Fe_{2/3}S + S = 2Fe_{3/4}S + FeS_2$$
$$4Fe_{3/4}S + S = 3Fe_{4/5}S + FeS_2$$
$$5Fe_{4/5}S + S = 4Fe_{5/6}S + FeS_2$$
$$6Fe_{5/6}S + S = 5Fe_{6/7}S + FeS_2$$

. . . .

or, in the general form

$$\frac{1}{x} Fe_{1-x}S + S = \frac{1-x}{x} Fe_{1-x/(1-x)}S + FeS_2. \qquad (53)$$

For the compound $Fe_{1-x/(1-x)}S$

$$\Delta G^0 = -\frac{23.32 + 16.52x}{1+x}$$

$$\Delta G^0 = -39.84 + 23.32\left[1 - \frac{0.58x^2}{(1+x)(1+2x)}\right]$$

Here, ΔG^0 proves to be negative throughout the entire range of changes in x [0, 1/3], since between these limits

$$|-39.84| > 23.32\left[1 - \frac{0.58x^2}{(1+x)(1+2x)}\right]$$

The reaction

$$Fe_2S_3 = FeS + FeS_2$$

proposed by Rodt [1916] probably does not hold in the range of investigated temperature and pressure, since $\Delta G^0 = +2.72$ kcal.

All the analyzed reactions attest to the fact that, whatever the intermediate sulfide that formed initially in the sediment, it may be changed to pyrite in the presence of elemental sulfur. If we assume that there is no reserve of elemental sulfur in the medium, but that the amount of sulfur may be determined from the equation $H_2S = S + 2H^. + 2e$, where the elemental sulfur being formed is immediately used in the formation of pyrite, then reaction (53) is written in the following form:

$$\frac{1}{x} Fe_{1-x}S + H_2S = \frac{1-x}{x} Fe_{1-x/(1-x)}S + 2H^. + 2FeS_2 + 2e$$

$$Eh = -0.216\left[1 + 1.365x\left(1 + \frac{1}{(1+x)}\right)\right]$$

$$-0.059\,pH - 0.029\,\lg a_{H_2S} \qquad (54)$$

When the hydrogen-sulfide-sulfide system determines potential, the oxidation-reduction conditions for the change to pyrite may be determined from Eq. (39). Having set these equations equal to each other, according to Eh, we find

$$x^2 + 3.214x + 1.214 = 0$$

From this it is seen that in the zone of positive Eh values, the equation has no solution. The deficiency of hydrogen sulfide for the formation of pyrite under conditions that the hydrogen-sulfide-sulfur system will determine potential may arise only in the region of negative values of x, i.e., only where the problem loses real meaning. Thus, thermodynamic calculations indicate that in recent marine sediments only two groups of iron minerals may be stable (we are not considering iron silicates such as chamosite or glauconite here): ferric-hydroxide compounds and sulfides. The formation of any particular forms of sulfides is determined by a whole series of factors. But all sulfides are formed in a rather high zone relative to the oxidation–reduction potential, and they may all be converted to the most stable sulfide form: pyrite.

The formation of magnetite and siderite from ferric hydroxide should prove possible only under more reducing conditions than those for the formation of sulfides. Therefore, in marine sediments, when the hydrogen-sulfide–sulfur system determines potential, these minerals cannot form. The formation of primary siderite is possible in marine sediments only when $P_{CO_2} > 10^{-2}$ atm. If the sulfate ion is not being reduced, siderite may form over a wider range and, according to P_{CO_2}, may characterize more strongly oxidizing or more strongly reducing conditions than those under which magnetite forms.

Before falling into a zone where the hydrogen-sulfide system determines potential, ferric hydroxide passes through the zone where hydrogen sulfide is more abundant than required by Eq. (39). It is natural that the discussed factors will be in force in this zone, but it is even clearer that the possibility of forming siderite diminishes, and sulfides begin to form under somewhat more strongly oxidizing conditions than stipulated by Eqs. (47) and (48).

Earlier we considered the facts of increase in oxidation–reduction potential and decrease in quantity of sulfide ion downward in the sediments, beginning from the bacteriologically active layer, in order to prove that in recent sediments, where sulfate-reducing bacteria are active, the hydrogen-sulfide–sulfur system determines the potential. This view has been confirmed by the nature of the authigenic minerals in modern marine basins. We shall pause in more detail on the question of what the decrease in hydrogen sulfide is connected with and what consequences the decrease in hydrogen-sulfide content may have on diagenesis.

Intense microbiological activity is characteristic of only a thin layer of muds 10-25 cm thick. This active layer may be at the surface of the muds if the natural waters are poorly ventilated, or it may lie at some depth from the surface of the muds. Below this layer, the activity of microorganisms declines sharply. This peculiarity in the distribution of microorganisms has long been studied by microbiologists and, in essence, its existence is not denied. We shall therefore dwell no longer on more details concerning the mere fact of distribution of microorganisms in muds.

There is no necessity of discussing the causes of this phenomenon, since it is purely a microbio-

logical problem, going beyond the frame of our present work and of our competency. However, the consequences of the decline in microbiological activity downward from the active layer may be considered (for this there is no need to be a specialist in microbiology): if the activity of microorganisms dies out, then, consequently, the reduction of elements must also die out, particularly the process of sulfate reduction. In the microbiologically active layer, a_{H_2S} is always maintained at some definite level, since the loss of hydrogen sulfide in the various processes of reduction and diffusion is compensated by the activity of sulfate-reducing microorganisms. Below the active layer the depletion of hydrogen sulfide begins to surpass the reduction of sulfate ions, which leads to a decline in hydrogen-sulfide content. Unfortunately, there is at present insufficient material to take into account all the means by which hydrogen sulfide is consumed. Some part of it apparently goes for the slow reduction of certain types of organic material. It is very likely that some part is removed because of diffusion. It may be stated with certainty that some hydrogen sulfide is consumed in diagenetic transformations of sulfides. Let us recall the process of pyrite formation. Equation (54), defining this process, may be written in the following form:

$$\lg a_{H_2S} = -rH_2 - 7.32 \left[1 + 1.365x \left(1 + \frac{1}{1+x}\right)\right]$$

where

$$rH_2 = \frac{Eh + 0.059pH}{0.029}.$$

In this form the equation indicates that at any particular values of Eh and pH the transformation to pyrite + residual sulfide from the $Fe_{1-x}S$ type of sulfide requires a certain amount of hydrogen sulfide. Apart from this, the transformation proceeds in the direction that always diminishes the content of the sulfide ion in solution. This may be proved by the calculations below.

The solubility product of each sulfide is determined from the reactions

$$Fe_{3/4}S = {}^{0}/_3 Fe^{··} + {}^{3}/_3 Fe^{···} + S''$$
$$Fe_{3/4}S = {}^{1}/_4 Fe^{··} + {}^{2}/_4 Fe^{···} + S''$$
$$Fe_{4/5}S = {}^{2}/_5 Fe^{··} + {}^{2}/_5 Fe^{···} + S''$$
$$Fe_{5/6}S = {}^{3}/_6 Fe^{··} + {}^{2}/_6 Fe^{···} + S''$$

. . . .

In the general form

$$Fe_{1-x}S = (1 - 3x) Fe^{··} + 2x Fe^{···} + S''$$

According to standard free energies, it may be computed that

$$a_{Fe^{..}}^{(1-3x)} \cdot a_{Fe^{...}}^{2x} \cdot a_{S''} = 10^{-\left(1.22+40.25x+\frac{17.10+29.21x}{1+2x}\right)}$$

For further computations let us consider some simple examples. The solubility of a pure compound is easily calculated from the solubility product. Actually, if we place some FeS in pure water, dissociation occurs into the ions $Fe^{..}$ and S''. The number of these ions in the solution proves to be $Fe^{..} \cdot S'' = SP$. Since we are speaking of pure substances, the equality $Fe^{..} = S''$ should hold. Therefore, the activity of each ion may be written as

$$a_{Fe^{..}} . a_{S^{.}} = \sqrt{SP}$$

When Fe_2S_3 is dissolved, the situation is more complex. The solubility of salt of this type is equal to twice the activity of the cation and three times the activity of the anion. That is,

$$2a = Fe^{..} \quad \text{and} \quad 3a = S''$$

Substituting the obtained values in the equation

$$Fe^{...^2} \cdot S''^{3} = SP$$

we find SP:

$$SP = 4a^2 \cdot 27a^3$$

In similar fashion, the solubility is determined for each case by the equation

$$a = \sqrt[(2-x)]{\frac{10^{-\left(1.22+40.25x+\frac{17.10+29.21x}{1+2x}\right)}}{(1-3x)^{(1-3x)} \cdot (2x)^{2x}}}$$

From this equation we may calculate that

$$a_{S''} = 10^{-\left[\frac{1.22+40.25x}{(2-x)} + \frac{17.10+29.21x}{(2-x)(1+2x)}\right]} \cdot (1-3x)^{-\frac{(1-3x)}{(2-x)}} \cdot (2x)^{-\frac{2x}{2-x}}$$

$$a_{Fe^{..}} = a_{S''} \cdot (1-3x)$$

$$a_{Fe^{...}} = a_{S''} \cdot 2x$$

The graph computed from these equations is shown in Fig. 47, from which it is seen that with progress of diagenesis the amount of pyrite increases, and, consequently, in the direction of the process from $x = 0$ to $x = 1/3$ the amount of sulfide ion progressively diminishes.

We have made all these calculations in order to show one cause of loss of hydrogen sulfide under conditions of its weakened production, but we do not maintain that this process is the dominant one. There are probably other means of hydrogen-sulfide consumption that lead to lowering of its content downward in the sediments. The very fact that the content of hydrogen sulfide declines conceals a certain danger for the trend of mineral transformations that began in the upper part of the muds of marine basins. This danger involves the fact that with consumption of hydrogen sulfide, the poise of the hydrogen-sulfide—sulfur system declines. Ultimately this process should lead to the loss of potential-determining status at some time by the hydrogen-sulfide—sulfur system and to its replacement by some other more highly poised system in the complex conditions of the system, which prevents further increase in the oxidation—reduction potential because of decline in a_{H_2S}. When the sulfate ion is

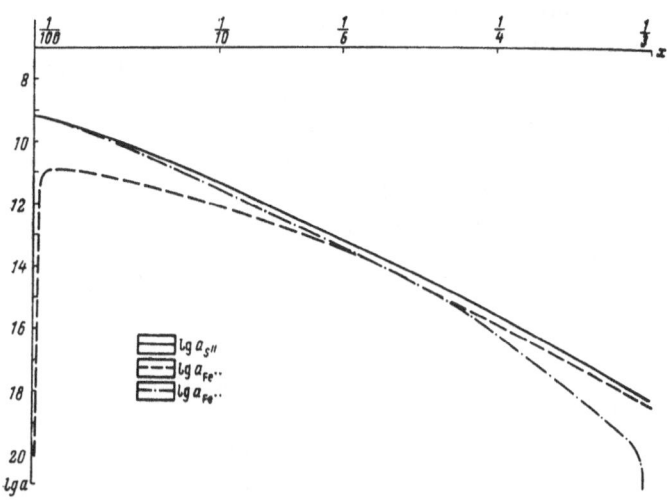

Fig. 47. Nature of the change in activities of ions in solution during transformation of iron sulfides in the presence of elemental sulfur.

not being reduced and elemental sulfur is being consumed in the formation of pyrite, a deficiency of elemental sulfur must result. The activity of sulfur cannot be taken as unity, which would invalidate Eq. (39). This equation then takes on the following form:

$$Eh = 0.142 + 0.029 \lg a_{S'} - 0.059 pH - 0.029 \lg a_{H_2S}$$

i.e., a_{H_2S} ceases being controlled merely by pH and Eh and becomes one of the independent variables. One of the first consequences of this process is suspension of the conversion of sulfide into pyrite. Under these conditions, some sulfide with the value $0 < x < 1/3$ occurs in paragenetic relation with pyrite. The nature of this sulfide will depend on the conditions under which the insufficiency of hydrogen sulfide appears relative to the oxidation–reduction potential. If at that time there has not occurred sufficient dehydration of the sediments, cessation of pyrite formation will not stop the conversion of $Fe_{1-x}S$ sulfides. This conversion will continue because of the necessity of forming compounds more stable under the developing conditions, adapted to the declining activity of the sulfide ion. The conversion will continue until the sediments consolidate and chemical reactions prove almost impossible. A graphic illustration of the nature of this conversion as a function of Eh and Σa_S at a pH of 8 may be seen in Fig. 48.

It would seem that in this situation, according to the nature of the residual sulfides, it might be possible to determine the approximate oxidation–reduction state of the sediments at the end of diagenesis. Actually this is impossible, since, along with the change of sulfides in the sediments, other mineral transformations occur, leading to partial or complete destruction of $Fe_{1-x}S$ sulfides. In particular, in connection with the progressive decline in hydrogen-sulfide content not compensated by rise in oxidation–reduction potential, conditions may arise in which magnetite and siderite form.

Let us pause for more details on these processes. Earlier we wrote the reaction for iron sulfide and magnetite paragenesis. We shall rewrite this reaction for acidic and alkalic conditions:

for $pH < 7$, $\frac{3}{1-x} Fe_{1-x}S + 4H_2O = Fe_3O_4 +$

$$+ \frac{3}{1-x} H_2S + 2 \frac{1-4x}{1-x} H^{\cdot} + 2 \frac{1-4x}{1-x} e$$

for $pH > 7$, $\frac{3}{1-x} Fe_{1-x}S + 4H_2O = Fe_3O_4 +$

$$+ \frac{3}{1-x} HS' + \left(8 - \frac{3}{1-x}\right) H^{\cdot} + 2 \frac{1-4x}{1-x} e$$

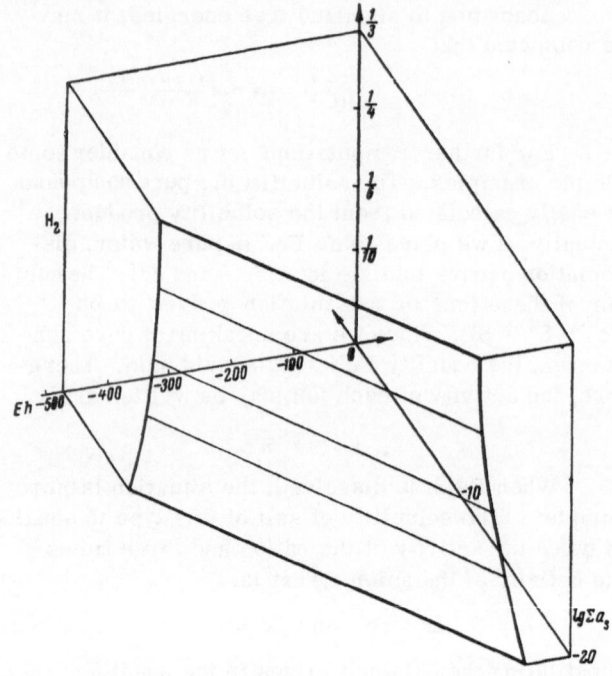

Fig. 48. Nature of the transformation of iron sulfides according to changes in Eh and Σa_S at a pH of 8.

The corresponding equations for these paragenetic states, calculated by the ordinary method, may be written in the following form:

for $pH < 7$, $Eh = 1.09 \frac{(1 + 1.59x)}{(1 - 4x)(1 + 2x)} - 0.11 \frac{(1-x)}{(1-4x)} +$

$$+ \frac{0.0885}{(1-4x)} \lg \sum a_S - 0.059 pH$$

for $pH > 7$, $Eh = 1.71 \frac{(1 + 1.74x)}{(1 - 4x)(1 + 2x)} - 0.11 \frac{(1-x)}{(1-4x)} +$

$$+ \frac{0.0885}{(1-4x)} \lg \sum a_S - 0.0295 \left(2 + \frac{3}{1-4x}\right) pH$$

Apart from the change to magnetite, when the oxidation-reduction potential is sufficiently high, iron sulfide may go directly to ferric hydroxide according to the reactions

for $pH < 7$, $\frac{3}{1-x} Fe_{1-x}S + 9H_2O = 3Fe(OH)_3 +$

$$+ \frac{3}{1-x} H_2S + 3 \frac{1-3x}{1-x} H^{\cdot} + 3 \frac{1-3x}{1-x} e$$

for $pH > 7$, $\frac{3}{1-x} Fe_{1-x}S + 9H_2O = 3Fe(OH)_3 +$

$$+ \frac{3}{1-x} HS' + 3 \frac{2-3x}{1-x} H^{\cdot} + 3 \frac{1-3x}{1-x} e$$

The following oxidation–reduction equations correspond to this paragenesis:

for $pH < 7$, $Eh = 0.73 \frac{(1 + 1.59x)}{(1 - 3x)(1 + 2x)} +$

$$+0.03\frac{(1-x)}{(1-3x)}+\frac{0.059}{(1-3x)}\lg\sum a_S-0.059pH$$

for $pH>7$, $Eh=1.14\frac{1+1.74x}{(1-3x)(1+2x)}+$

$$+0.03\frac{(1-x)}{(1-3x)}+\frac{0.059}{(1-3x)}\lg\sum a_S-0.059\left(1+\frac{1}{1-3x}\right)pH$$

For acidic conditions, when the activity of the ferrous ion may be appreciable, it is necessary to use the reaction

$$(1-x)\,Fe^{..}+H_2S=Fe_{1-x}S+2H^{.}+2xe$$

to which the following equation corresponds:

$$Eh=\frac{0.08+0.3x}{(1+2x)\,x}-0.44-\frac{0.059}{x}\,pH-$$

$$-\frac{0.029}{x}\lg\sum a_S-0.029\frac{1-x}{x}\lg a_{Fe^{..}}\qquad(55)$$

Specific diagrams for $a_{Fe^{..}}=10^{-3}$ mole, illustrated in Figs. 49–52, were plotted by means of these equations. In examining the diagrams it is not difficult to conclude that the mineral most sensitive to increase in Eh and decrease in total activity of hydrogen sulfide proves to be the FeS sulfide. The least sensitive in these relations is the Fe_2S_3 sulfide. Thus, according to the specific aspects of

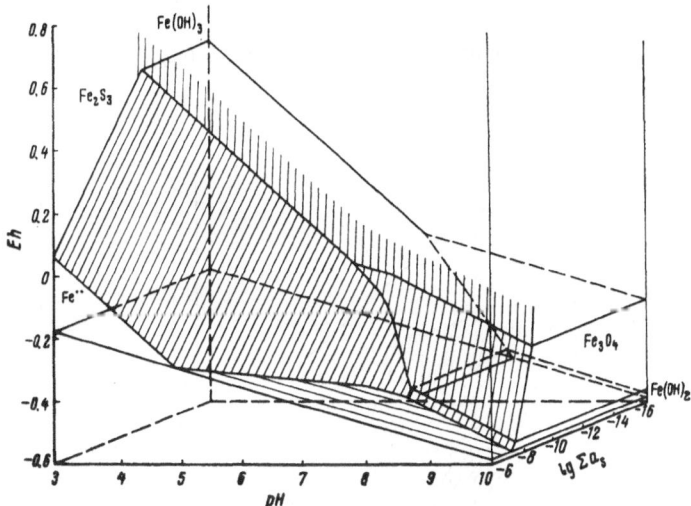

Fig. 49. Stability diagram for iron minerals ($a_{Fe^{..}}=10^{-3}$).

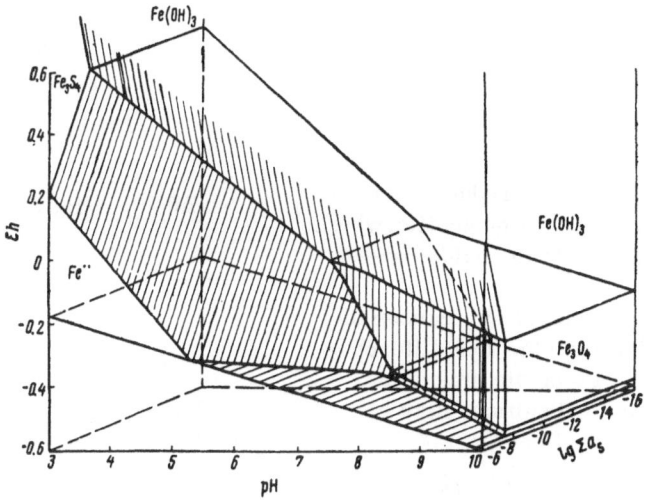

Fig. 50. Stability diagram for iron minerals ($a_{Fe^{..}}=10^{-3}$).

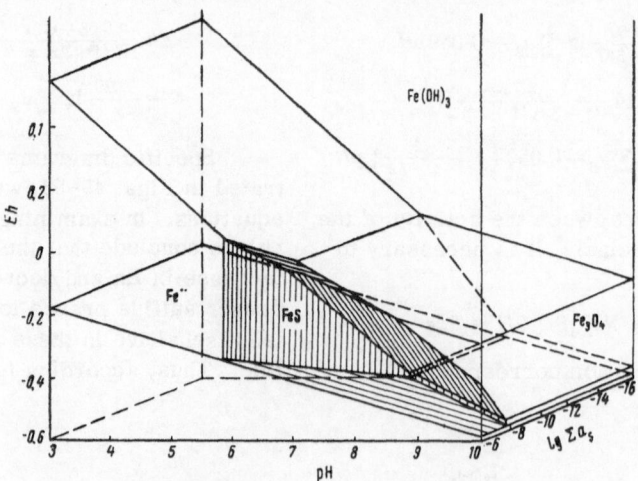

Fig. 51. Stability diagram for iron minerals ($a_{Fe^{..}} = 10^{-3}$).

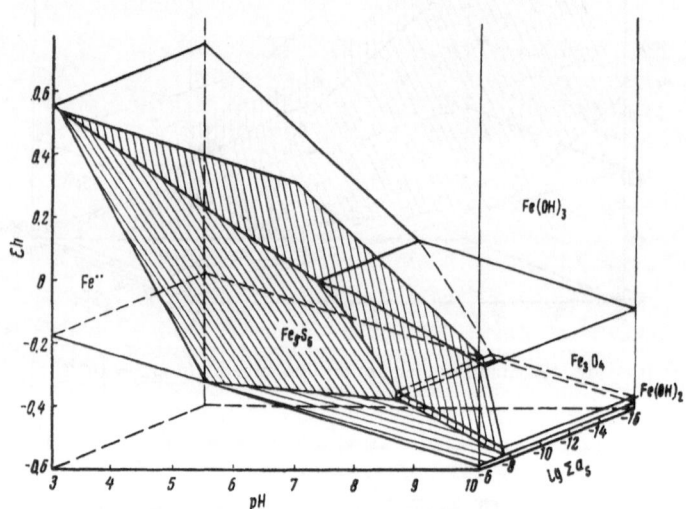

Fig. 52. Stability diagram for iron minerals ($a_{Fe^{..}} = 10^{-3}$).

the diagenetic processes, a certain reduced part of the series of sulfides is changed to magnetite, which will be paragenetically related to the oxide part of the sulfide series.

    If, in addition to hydrogen sulfide, carbon dioxide is also present in the system, then different variants for the formation of siderite may arise. The paragenesis of siderite and sulfides may be described by the following reactions:

$$(1-x)\ FeCO_3 + H_2S = Fe_{1-x}S + (1-x)\ CO_2$$

$$+ (1-x)\ H_2O + 2xH^{\cdot} + 2xe.$$

$$\Delta G^0_{reaction} = 16.65 - 10.11x - \frac{23.32 + 39.84x}{(1+2x)}$$

$$Eh = \frac{0.361}{x} - \frac{0.506 + 0.864x}{(1+2x)} - 0.219 - 0.059\,pH +$$

$$+ \frac{0.059\,(1-x)}{2x}\ \lg P_{CO_2} - \frac{0.029}{x}\ \lg a_{H_2S}$$

    This equation will characterize the acidic zone. For alkaline conditions we shall use the substitute

$$a_{H^{\cdot}} \cdot a_{HS^{\cdot}} = 10^{-7} \cdot a_{H_2S}$$

As a result we obtain two equations:

for $pH < 7$,

$$Eh = \frac{0.361}{x}\frac{0.506+0.864x}{(1+2x)} - 0.219 - 0.059\,pH +$$

$$+ \frac{0.059\,(1-x)}{2x}\lg P_{CO_2} - \frac{0.029}{x}\lg \sum a_S$$

and, for $pH > 7$,

$$Eh = \frac{0.155}{x} - \frac{0.506+0.864x}{(1+2x)} - 0.219 +$$

$$+ 0.029\left(\frac{1}{x}-2\right)pH + \frac{0.059\,(1-x)}{2x}\lg P_{CO_2} - \frac{0.029}{x}\lg \sum a_S$$

In addition, for the acidic zone, where high activity of the ferrous ion is acceptable, it is necessary to use Eqs. (26), (28), and (55).

In the system of equations used for plotting the graph, six independent variables are introduced, considerably complicating the possibility of graphical representation. One way out is to construct specific diagrams for several sulfides with some fixed activity of the ferrous ion, as was done for magnetite. In this case the question is still complicated by the choice of a value for one variable: P_{CO_2}.

In order to encompass all possible cases with a smaller number of graphs, it is necessary to start from those two points of partial pressure of carbon dioxide which lead to fundamental rearrangement of the mineral-forming process: specifically $P_{CO_2} = 10^{-2}$ atm, and $P_{CO_2} = 2.55 \cdot 10^{3}$ atm. At $P_{CO_2} > 10^{-2}$ atm the first mineral proves to be siderite. Sulfides are formed under more strongly reducing conditions. At $P_{CO_2} < 10^{-2}$ atm siderite is possible only as a secondary mineral. At $P_{CO_2} < 2{,}55 \cdot 10^{-3}$ atm magnetite will characterize the more oxidizing conditions of the medium as contrasted with siderite. At $P_{CO_2} > 2.55 \cdot 10^{-3}$ the reverse will be true. Therefore, to obtain a generalized concept of the paragenetic relations among magnetite, siderite, $Fe_{1-x}S$ sulfides, and pyrite under conditions in which $Fe_{1-x}S$ sulfides form first and all the other minerals are secondary, two series of diagrams are necessary: 1) $P_{CO_2} = 10^{-2}$ atm, and $P_{CO_2} = 10^{-4}$ atm.

for $pH < 7$,

$$Eh = \frac{0.30}{x} - 0.16\frac{0.506+0.864x}{(1+2x)\,x} - 0.059\,pH - \frac{0.029}{x}\lg \sum u_S$$

for $pH > 7$,

$$Eh = \frac{0.10}{x} - 0.16 - \frac{0.506+0.864x}{(1+2x)\,x} + 0.029\left(\frac{1}{x}-2\right)pH - \frac{0.029}{x}\lg \sum a_S$$

$\left.\right\} P_{CO_2} = 10^{-2}$ atm

for $pH < 7$,

$$Eh = \frac{0.24}{x} - 0.1 - \frac{0.506+0.864x}{(1+2x)\,x} - 0.059\,pH - \frac{0.029}{x}\lg \sum a_S$$

for $pH > 7$,

$$Eh = \frac{0.04}{x} - 0.1 - \frac{0.506+0.864x}{(1+2x)\,x} + 0.029\left(\frac{1}{x}-2\right)pH - \frac{0.029}{x}\lg \sum a_S$$

$\left.\right\} P_{CO_2} = 10^{-4}$ atm

These equations serve as the basis for the graphs in Figs. 53−55.

As seen from these figures, we note the same pattern of sequential replacement of sulfide by siderite as noted in regard to magnetite. The difference lies in a somewhat smaller increase in Eh, in regard to siderite, than the decrease in activity of the sulfide ion. Nevertheless, we clearly see that the decrease in sulfide ion uncompensated by Eh must lead to replacement of the reduced part of the sulfide by siderite.

Indicators of Oxidation − Reduction Conditions under Which Sedimentary Rocks Form

It is now possible to approach the evaluation of iron minerals, using them as indicators of the oxidation−reduction conditions prevailing during the formation and diagenesis of sediments. Analysis of the possibilities of formation of any particular mineral forces us first of all to assign an important place to the gaseous components of the medium. The formation of minerals examined in this book depends on, apart from Eh and pH, the gas that participates in the mineral-forming process: oxygen, carbon dioxide, or hydrogen sulfide. Without the participation of carbon dioxide and hydrogen sulfide, depending on the oxidation−reduction conditions, either ferric hydroxide or magnetite might form. The participation of CO_2 leads to the appearance of siderite, and the effect of H_2S not only causes the appearance of a new mineral form, the sulfides, but also a fundamental rearrangement in the sequence of mineral formation. Oxygen and hydrogen sulfide

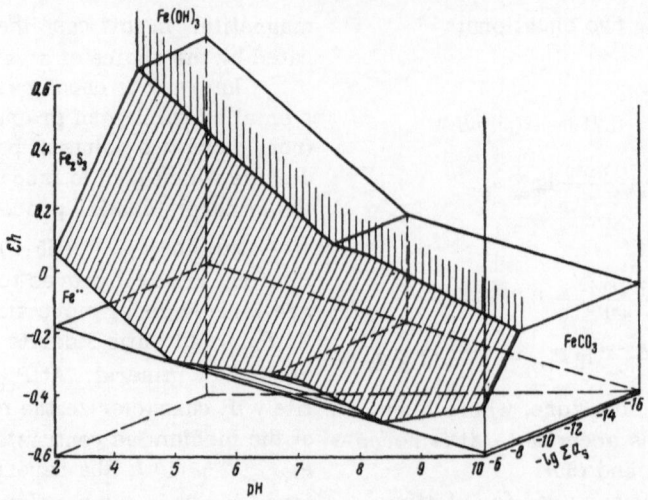

Fig. 53. Stability diagram for ferric hydroxide, Fe_2S_3 sulfide, and siderite at $a_{Fe^{..}} = 10^{-3}$ mole; $P_{CO_2} = 10^{-2}$ atm.

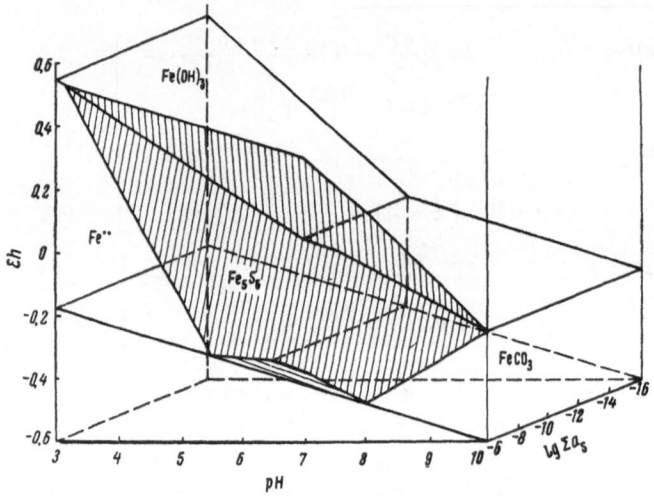

Fig. 54. Stability diagram ferric hydroxide, Fe_5S_6 sulfide, and siderite at $a_{Fe^{..}} = 10^{-3}$ mole; $P_{CO_2} = 10^{-2}$ atm.

completely (if the system is potential-determining)
or partly (if the system is not potential-determining)
determine the oxidation–reduction state and the
acid–alkaline conditions of the environment, whereas P_{CO_2} does not depend on these parameters. In
connection with the effect of P_{CO_2} on the sequence
of mineral formation, it would be important to know
the limits, even if approximate, between which P_{CO_2}
might have changed during the formational history
of the earth's sedimentary shell.

    For plotting the stability fields of siderite by
the equations derived in Chap. 3, it is possible to

assign arbitrary values of P_{CO_2} and to construct
the diagrams in Eh–pH–$a_{Fe^{..}}$ coordinates. In particular, we might use the data on P_{CO_2} used in constructing diagrams by Krumbein and Garrels [1960],
Huber [1960], or Krauskopf. The values used by
these authors, however, represent but a random
event in the history of the earth: a basin in the
present-day stage of development. Unfortunately,
investigators who have used the diagrams of Krumbein and Garrels have given little attention to these
characteristic aspects. A random value of P_{CO_2} or
ΣCO_2 is generalized over the entire history of exis-

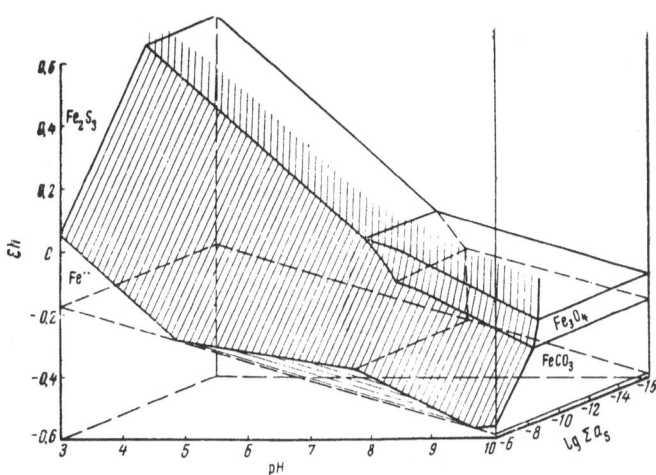

Fig. 55. Stability diagram for iron minerals at $a_{Fe^{..}} = 10^{-3}$ mole; $P_{CO_2} = 10^{-4}$ atm.

tence of the earth's sedimentary shell, and siderite is treated as an indicator, appearing at a definite Eh value. In particular, opinions have been expressed that the formation of siderite in sedimentary rocks begins at a potential of 0 mV [Teodorovich, 1956] or 0.12 mV [Tochilim, 1956].

As a maximum limit for the CO_2 content we may adopt $P_{CO_2} = 1$ atm, since it is doubtful that the general atmospheric pressure exceeded this value in the past. It is a more complex problem to set a definite lower limit. As a first approximation, however, we may find some basis for setting this limit. For this it is necessary to examine the paragenesis of magnesite and brucite in sedimentary rocks. These minerals may be found in paragenetic relationships under certain relations of the CO_3'' and OH' ions, as seen from the following:

$$a_{Mg^{..}} \cdot a_{CO_3''} = MgCO_3, \quad SP = 1 \cdot 10^{-5} \,^{\dagger}$$

$$a_{Mg^{..}} \cdot a_{(OH)'}^2 = Mg(OH)_2 \quad SP = 8.9 \cdot 10^{-12} \,^{\ddagger}$$

Whence

$$\frac{a_{Mg^{..}} \cdot a_{CO_3''}}{a_{Mg^{..}} \cdot a_{(OH)'}^2} = 1.12 \cdot 10^{-6}$$

Since both minerals contain the same ion, $Mg^{..}$, for their simultaneous precipitation we may write

$$\frac{a_{CO_3''}}{a_{(OH)'}^2} = 1.12 \cdot 10^{6} \tag{56}$$

In examining the system $Fe(OH)_3 - Fe_3O_4 - Fe(OH)_2$ we have already used the expression $a_{H^{.}} \cdot a_{(OH)'} = 1.27 \cdot 10^{-14}$. Whence

$$a_{(OH)'}^2 = \frac{1.6 \cdot 10^{-28}}{a_{H^{.}}^2}$$

Substituting the obtained values in expression (56), we find

$$a_{CO_3''} = 1.8 \cdot 10^{-22} \cdot a_{H^{.}}^{-2}$$

In accordance with the conditions of paragenesis with the inequality

$$a_{CO_3''} > 1.8 \cdot 10^{-22} \cdot a_{H^{.}}^{-2}$$

in the sediment, only magnetite will form.

When

$$a_{CO_3''} < 1.8 \cdot 10^{-22} \cdot a_{H^{.}}^{-2}$$

brucite begins to form. In replacing a_{CO_2} by P_{CO_2} from Eq. (21) for the conditions of brucite occurrence, we find

$$P_{CO_2} \leqslant 2.4 \cdot 10^{-4}$$

Thus, the change from nesquehonite to brucite does not depend on pH, but is determined merely by the partial pressure of carbon dioxide. Strakhov [1954] cited data on the partial pressure of carbon dioxide in the modern atmosphere, expressed by the numbers $3.7 \cdot 10^{-4}$ atm (Johnson) and $3.8 \cdot 10^{-4}$ atm (Klein). Since brucite is an extremely rare mineral in sedimentary rocks and modern soda lakes, the cited data on partial pressure of carbon dioxide must be considered the lowest that might have existed at some time in the development of the earth's sedimentary shell. Therefore, rounding it off,

---

† The solubility product of $MgCO_3$ is $8 \cdot 10^{-9}$, but, since this compound precipitates in basins as nesquehonite $(MgCO_3 \cdot 3H_2O)$, we give here the solubility product of the latter. The SP value is taken from Nadeinskii [1956].

‡ The SP value is taken from Latimer [1952].

we may take the value of $P_{CO_2} = 1 \cdot 10^{-4}$ atm as the minimal limit of carbon dioxide in the history of the earth.

The investigated conditions represent the ideal variant, not only because arbitrary data are used here, but also because, in connecting the partial pressure of the gas with solubility, we did not consider variation in salinity of the basins, which naturally affects the solubility of a gas. To evaluate the effect of salinity, we use the equation of Sechenov:

$$\lg \frac{S}{S'} = Km$$

where S and S' represent the solubility of the gas in pure and mineralized water, respectively; K is the coefficient of Sechenov's equation, depending on the kind of gas and solvent, temperature, pressure, and the selected unit of salt concentration; and m is the salt concentration in g-eq/liter.

Since the logarithm is of a ratio, the solubility of the gas may be expressed in arbitrary units. We shall use solubility in moles per liter.

Sechenov's equation may be rewritten in the form

$$\lg \sum CO_2 = K_m + \lg \sum{}' CO_2$$

and we may substitute the value of $\lg \Sigma CO_2$ obtained in any of Eqs. (30)−(32). As an example we may consider this procedure in Eq. (30).

After substitution

$$Eh = 0.55 + 0.059K_m = 0.059pH + 0.059 \lg \sum{}' CO_2$$

If we change the partial pressure so that $\Sigma' CO_2 = \Sigma CO_2$, this somewhat displaces the potential toward the value Eh':

$$Eh' = 0.55 + 0.059K_m - 0.059pH + 0.059 \lg \sum CO_2 \quad (57)$$

Computing Eq. (30) from Eq. (57), we find

$$\Delta Eh = 0.059K_m$$

Mishina, Avdeeva, and Bozhovskaya [1961] determined that, up to a salinity of 2 g-eq/liter, K = 0.101, and from a salinity of 2 to a salinity of 5.4, K = 0.096 g-eq/liter. Hence, we obtain, with accuracy to three places, for any salinity

$$\Delta Eh = 0.006M$$

With these data we have plotted a graph (Fig. 56) of the change in ΔEh at the beginning of siderite formation versus salinity at a particular partial pressure of carbon dioxide. From this graph we may see that with increase of salinity the region of siderite formation should diminish. However,

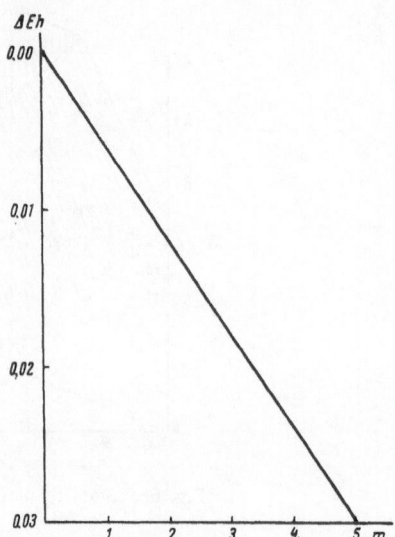

Fig. 56. Amount of error introduced by salinity of a basin in determining oxidation−reduction potential at the beginning of siderite precipitation.

even at rather high salinities, i.e., those at which common salt begins to precipitate, ΔEh does not exceed 30 mV, within the accuracy of ordinary measurements. Therefore, diagrams plotted for pure solutions will be effective for saline water as well. Thus, when analyzing the formation of minerals for waters of any salinity, we shall restrict P_{CO_2} to the range of 1 atm to 10^{-4} atm.

As we have already pointed out, the appearance of hydrogen sulfide in the system is accompanied not only by the appearance of sulfides but also by a fundamental rearrangement of the sequence of mineral formation. In order to exclude the appearance of authigenic syngenetic siderite and magnetite, it is sufficient that $\Sigma a_S = 10^{-10}$ mole/liter, a negligibly small value. In this connection, it is advisable when selecting conditions for the formation of minerals to choose two types of environments:

1. mineral formation in the absence of free hydrogen sulfide;
2. mineral formation with the participation of hydrogen sulfide.

Mineral formation under conditions of no free hydrogen sulfide is apparently characteristic of fresh-water basins, in which there is no sulfide ion and in which the processes of microbiological reduction of sulfate are suppressed. Unfortunately, data on the distribution of oxidation−reduction potential vertically through sediments in basins of this type are not available. It must be thought that the general pattern prevails here: the lowest oxidation−reduction potential is characteristic of the layer with intense activity of microorganism, and downward from this there should occur a slight rise in the potential.

In support of this we may cite the conclusions of Kuznetsov [1952], who investigated lakes both with and without the manifestation of sulfate reduction. For the formation of minerals, the first type is characterized by no such role of the oxidation−reduction potential as observed in the second. In the second type the minerals may form in the following sequence: ferric hydroxide, magnetite, siderite or ferric hydroxide, siderite, depending on P_{CO_2}. Each of these minerals may be syngenetic.

Mineral formation of the second type is characteristic of basins the waters of which contain sulfate ions capable of being reduced. In this category we should include all marine and oceanic basins, and also saline lakes. The appearance of hydrogen sulfide in a medium, the product of sulfate reduction, has two consequences: first, it adds an ion that is capable of producing a solid phase with ferrous and ferric ions, and second, arising by microbiological means without being in equilibrium with the environment, hydrogen sulfide itself may initiate changes in the oxidation−reduction conditions. Therefore, according to whether hydrogen sulfide determines the potential of the system, the process of mineral formation will vary. If the hydrogen-sulfide system determines potential, two sequences of authigenic syngenetic minerals are theoretically possible: $P_{CO_2} > 10^{-2}$ atm, yielding ferric hydroxide, siderite, and iron sulfides; and $P_{CO_2} < 10^{-2}$ atm, yielding ferric hydroxide and iron sulfides.

The ferric-hydroxide−iron sulfide transition occurs at a higher oxidation−reduction level than the corresponding boundaries for the beginning of magnetite and siderite formation. Therefore, before hydrogen sulfide is present in amounts to determine the oxidation−reduction potential, and P_{CO_2} remains less than 10^{-2} atm, magnetite and siderite will not form. When the hydrogen-sulfide system does not determine potential, the formation of minerals is controlled by the relations of the hydrogen-sulfide system to the potential-controlling system. If a system stronger than the hydrogen-sulfide−sulfur system determines potential, hydrogen sulfide will most likely be in excess over the amount necessary for producing an oxidation−reduction potential for reaction (39).

In this case the nature of the mineral-forming processes remains as before. The difference is expressed in the greater restriction in the possible formation of siderite and, perhaps, its complete exclusion even at $P_{CO_2} = 1$ atm. The boundary of the transition from ferric hydroxide to sulfides is shifted toward the more highly oxidizing zone as compared with the boundary determined for hydrogen sulfide as the potential-determining system. A weaker system may become potential-determining when the activity of the sulfide ion proves to be insufficient according to the theoretical value computed from Eq. (39). As a result, sulfides begin to change either to siderite or to magnetite. If the consumed carbon dioxide is not replenished in the change of sulfides to siderite, magnetite may begin to form at a certain stage.

The change in role of the hydrogen-sulfide system in the sequence just discussed is characteristic of recent marine sediments. In places where microorganisms are very active, the oxidation−reduction potential exhibits a sharp peak toward negative values. If the water of the basin is but weakly ventilated and the sediments are very fine grained, negative values of the oxidation−reduction potential are observed at the very surface of the sediments. Where oxygen reaches the sediments, the layer of greatest reduction capacity is below the surface of the sediments a certain distance. The layer of greatest reducing capacity is at the same time the layer that generates hydrogen sulfide in greatest measure and the layer where hydrogen sulfide forms the potential-controlling system. Above and below this layer, hydrogen sulfide loses its role as the potential-determining system. Upward hydrogen sulfide proves to be in excess as compared with the theoretically computed amount from Eq. (39), and below it is deficient.

Throughout this entire work we have devoted considerable attention to an examination of the role of microbiological factors that have had a substantial influence on the formation of authigenic minerals, but in these discussions we have not touched at all on the energy sources for bacterial activity, at least not for heterotrophic microorganisms. This oversight was intentional, in order that the investigated questions, complex enough in themselves, should not be complicated by still other factors. Now, after investigating the principal aspects of the conditions for the formation of minerals in connection with the change in oxidation−reduction conditions, after supplying a general picture of mineral formation, it is necessary to evaluate the organic material in all the investigated processes.

Organic matter plays a tremendous role in changing the oxidation−reduction conditions, but no direct effect on the mineral-forming processes has been detected. With organic activity, a large part of the mineral-forming process still proceeds in the ordinary way. Everything depends on the nature of the microorganisms, the type of activity they manifest, the ionic composition, and the gaseous components of the medium. Connecting the mineral-forming processes with the amount of organic matter alone is to oversimplify the nature of the phenomenon.

Autotrophic anaerobic microorganisms, in manifesting their activities, do not directly use organic matter, but they may prepare the appropriate oxidation–reduction environment in a definite manner. Organic matter leaves its mark only when the part of it capable of reaction is insufficient for converting to the reduced form all the elements susceptible to reduction. Under these conditions, one may observe this phenomenon when, for example, the process of siderite formation (or sulfide formation) begins but does not go to completion, resulting in the occurrence of siderite (or pyrite) in the rock along with iron hydroxides. In this situation, organic matter determines the lack of reaction capacity.

The potential of those deposits should be indicated by a reduced mineral: siderite, pyrite, or magnetite. This potential will be constant regardless of the $Fe^{\cdot\cdot}/Fe^{\cdot\cdot\cdot}$ ratio obtained during analysis. The ratio of ferrous to ferric iron as an indicator of the oxidation–reduction conditions loses meaning even when organic matter is present in sufficient quantities, since the different ratios of ferric and ferrous ions may be due to individual forms of sulfides remaining after diagenetic transformations and also to other diagenetic minerals (such as Fe_3O_4).

As becomes clear from our discussion, the method of determining oxidation–reduction potential from residual organic matter is imperfect. Here lies the cause of all those disagreements concerning the determination of oxidation–reduction conditions we now have, and which we described in the first chapter.

Let us investigate what it is that iron minerals in a sedimentary rock may indicate.

1. In a rock containing only ferric hydroxide compounds. It may be stated with assurance that the environment in which the minerals formed was bounded by Eh > 0.48−0.059pH. In this case it is not clear what kind of basin the rock formed in, i.e., whether the basin contained the sulfate ion or not. Quaternary loams represent an example.

2. Siderite is present along with ferric hydroxides. Deposits were formed in a basin of the first type, i.e., in a basin in which the water contained no sulfate ion or the reduction of sulfate was suppressed. Conditions of formation of the sediments lay in the range (0.23−0.59pH) < Eh ≤ (0.47−0.059pH).

The limits were computed according to the maximum $P_{CO_2} = 1$ atm and minimum $P_{CO_2} = 10^{-4}$ atm. Microbiological processes could not manifest themselves in full measure either because of restricted energy reserves necessary for activity of the microorganisms or because of the accumulation of the products of microbiological catabolism, or, possibly, because of still other causes. If the principal cause was the shortage of organic matter, then, during analysis of the rock, residual organic matter corresponded (stochastically, but not functionally) with the $Fe^{\cdot\cdot}/Fe^{\cdot\cdot\cdot}$ ratio determined in a 5% (or 10%) hydrochloric-acid extract. This does not mean, however, that the oxidation–reduction potential is also determined by the ratio of the forms of iron. On the contrary, the joint occurrence of siderite and ferric hydroxide indicates that if P_{CO_2} remained constant, then the oxidation–reduction potential also remained constant, holding at some level within the indicated limits.

The joint discovery of siderite and ferric hydroxides must be evaluated very carefully. In connection with the low quantity of energy reserves necessary for microorganism activity, there is strong indication that the distribution of microorganisms is irregular in different parts of the mud. Therefore, the appearance of siderite may be due to local development of reduction processes around some organic center. The medium is characterized by a distinctive "mosaic equilibrium" and by the same mosaic distribution of corresponding oxidation–reduction conditions. This is precisely the character of the Kerch iron ores, in which siderite formation was mostly confined to wood fragments.

3. Rock represented by authigenic syngenetic siderite. It may be stated that for such rocks Eh ≤ 0.47−0.059pH and that the rocks belong to basins of the first type.

4. Only magnetite is present in the sedimentary rock. Such rocks are characterized by (0.23−0.059pH) < Eh ≤ (0.31−0.059pH), with P_{CO_2} no greater than 10^{-4} atm. Basins are of the first type.

5. Joint presence of magnetite and ferric hydroxides, similar to the presence of siderite and ferric hydroxides. For these sediments, Eh = 0.31−0.059pH with P_{CO_2} no greater than 10^{-4} atm if, finally, such a paragenesis is not affected by "mosaic equilibrium." This case also characterizes basins of the first type.

6. Siderite (first of the minerals that formed) and magnetite (second mineral that formed) characterize the sequential removal of CO_2 during mineral formation with transition to the critical boundary $P_{CO_2} = 3.55 \cdot 10^{-3}$ atm. The conditions for this paragenesis are characterized by Eh < 0.47—0.059pH with subsequent shift to Eh ≤ 0.31—0.059pH.

Magnetite (first mineral to form) and siderite (second mineral to form) reflect sequential decline of the oxidation—reduction potential throughout diagenesis at $P_{CO_2} < 3.55 \cdot 10^{-3}$ atm, Eh < 0.31—0.059pH. This paragenesis also characterizes basins of the first type.

Basins of the second type, in which sulfates are being reduced, are characterized by a more complex pattern of mineral formation. It is important to know in what part of the sediments the microbiologically active layer is located. According to the hydrodynamical conditions of the basin, the grain-size characteristics of the constituent sediments, the amount and nutriment value of the organic matter in the sediments, the rate of sedimentary accumulation, and other factors, sulfate reduction may occur in the surface layer of the muds but may also begin at some depth below the surface. The range of distances of the sulfate-reducing layer from the surface of the mud may be rather large. For example, in some parts of the Indian Ocean the layer of intense sulfate reduction lies at a depth of 11 m [Zheleznova and Shishkina, 1964]. Burial of the layer at some depth below the surface of the sediments is fostered by the invariable presence of hydrogen sulfide above this layer in amounts exceeding that specified in Eq. (39). As a result, sulfides begin to form under conditions near Eh = 0.48—0.059pH. The possibilities of sulfide formation are restricted by the oxidized part of the series. An abundance of elemental sulfur leads to a change of sulfides to pyrite, and in this process the mineral transformation proceeds to conclusion. The probability of sulfide formation in such sediments during diagenesis proves to be very low. Hence the following category of sediments.

7. Rock characterized by the presence of pyrite and the absence of other mineral forms of iron. This indicates that the rock belongs to the second type of basin. Hydrogen-sulfide contamination in all probability does not extend beyond the sediments. The sulfides began to form at Eh = 0.48—0.059pH. During further diagenesis the oxidation—reduction potential might decline even lower, but the sedimentary rock bears no traces of this. If we compare this value of the oxidation—reduction potential with any value characterizing the transition from hydroxides in the first type of sediments, it then appears that in all variants the formation of sulfides, going to pyrite, takes place under conditions more strongly oxidizing than the formation of other mineral forms of iron (except, of course, ferric hydroxide). Despite the general hydrogen-sulfide contamination, in sediments of the Black Sea one encounters sporadic zones where hydrogen sulfide does not occur above the surface of the muds on the floor. Such zones belong also in the described category of sediments. A. A. Lebedintsev turned his attention to the fact that in some places, where the mud gives off an odor of hydrogen sulfide, the mud is grayish green and bluish gray, not black because of the presence of hydrotroilite. On this point Isachenko [1914] expressed the following thought: "I have no data to explain why the mud does not become black where hydrogen sulfide and iron salts were present, but it seems to me that our attention must be focused on the fact that where iron sulfide forms and accumulates and where hydrogen sulfide slowly continues to form but oxygen dissolved in water has access to the iron sulfide, we also observe the conversion of this iron sulfide to pyrite: $FeS + H_2S + O - FeS_2 + H_2O$."

This fact indicates that, for a solution to the problem of the nature of a basin, it is not enough to study a single section, but it is necessary to make a three-dimensional survey of the sequence. In the general case, the layer of microbiological activity may be found at a certain depth below the surface of the mud. If the layer is buried to sufficient depth and the sediments accumulate at a slow rate, concretions of iron hydroxides may then form in the upper part of the mud.

The indicated method of mineral formation from the viewpoint of reducing processes becomes posssible when the activity of hydrogen sulfide proves to be insufficient to satisfy the solubility product of the sulfides ($a_{S^-} < 10^{-13}$ mole/liter). Negligibly small concentrations change the ferric ions to ferrous ions, which migrate in different directions. Descending, ferrous ions are bound with sulfides, but migrating upward, where the oxidation—reduction potential is above the boundary line of $Fe^{\cdot\cdot} - Fe(OH)_3$, the ion is oxidized, giving rise to a different kind of concretion.

With deposition of sediments, all these horizons are gradually shifted, and progressive reworking of previously formed compounds takes place in the sediments. Therefore, we find sediments of only the third stage, bearing pyrite, in sedimentary rocks. Sediments in the central part of the Indian Ocean are an example of modern sedi-

ments passing through such stages. A comparison
may be merely qualitative, since the pressure and
temperature on the floor somewhat alter the parameters we calculated for P = 1 atm and t = 25°C.
As an example of shallow-water sediments we may
refer to the classic section of the Galaikovtsy horizon of middle Paleozoic rocks in the middle Dniester
region, which the author once investigated [Stash-
chuk, 1958]. The pattern of changes in oxidation–
reduction conditions in this type of sediment is
shown schematically in Fig. 57.

8. Pyrite and diagenetic siderite are present
in the rock. Siderite here is a later mineral than
the pyrite. This paragenesis attests to weak venti-
lation of the sediments. It is therefore found most
frequently in very fine-grained sediments. Initially
a different type of sulfide was formed in the sediments, some of which was subsequently changed to
pyrite. With advance of diagenesis, when sulfur
ceased being the potential-determining system,
some of the sulfides were converted to siderite.
Sulfides began to form at Eh ≤ 0.48−0.059pH. After
lowering of the oxidation−reduction potential in the
microbiologically active layer, a certain increase
in Eh must have taken place, but the condition
Eh ≤ 0.48−0.059pH was preserved (Fig. 58). Judging from the description of Yagofarov and Gorelova
[1962], these conditions were characteristic of the
period when the Lower Carboniferous rocks of the
Kama and Middle Volga region formed. In general,

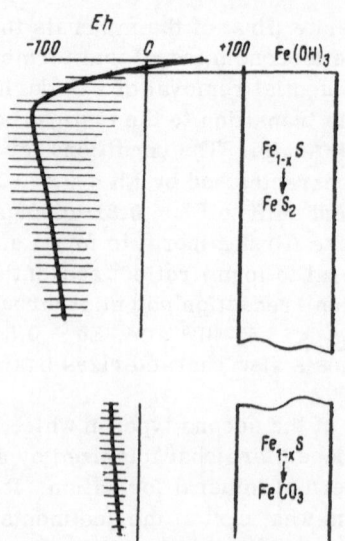

Fig. 58. Schematic picture of the
change in oxidation−reduction potential during the formation of a sedimentary rock in which the paragenetic pair
pyrite–siderite were observed.

this category of rocks, with nodules of pyrite and
bun-shaped masses of siderite, is the most widespread.

9. The following sequence of minerals is noted in this rock: $Fe_{1-x}S$ sulfides, pyrite, siderite,
and magnetite. The situation was as in the preceding, with this difference: during diagenesis P_{CO_2}
must have been below $3.55 \cdot 10^{-3}$ atm. Toward the
end of diagenetic transformations, the oxidation−
reduction potential could not have been above the
limit Eh = 0.31−0.059pH. An example of such rocks
is found in certain horizons of the Maikopian Series
where it is exposed, particularly on the northern
part of the Kerch Peninsula.

10. The following sequence of minerals was
detected in studying this rock: $Fe_{1-x}S$ sulfides, pyrite, and magnetite. This rock represents the low
limit of CO_2 content. The paragenesis of the sediments reflects the limits (0.23−0.059pH) ≤ Eh ≤
(0.31−0.059pH) (Fig. 59).

This category of sediments probably has
slight distribution. The indicated sequence of minerals has been observed by the author in recent sediments of the Sivash [Stashchuk et al., 1964].

11. If, lastly, only pyrite and $Fe_{1-x}S$ sulfide
are present in the rock, then, during diagenesis,
$P_{CO_2} < 3.55 \cdot 10^{-3}$, $^{-3}$, and Eh was within the limits
(0.31−0.059pH) < Eh ≤ (0.48−0.059pH).

As seen from the investigated examples, the
oxidation−reduction potential during sedimentation

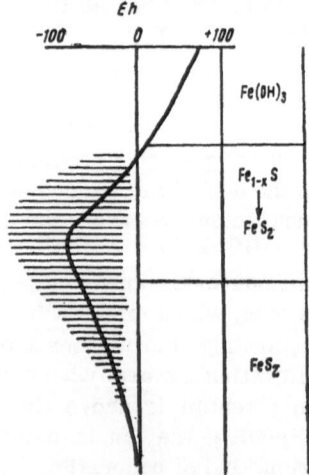

Fig. 57. Schematic picture of the
change in oxidation−reduction potential during the formation of sedimentary rock in which pyrite has
been discovered as the only authigenic mineral. Lined area represents possible variations in Eh.

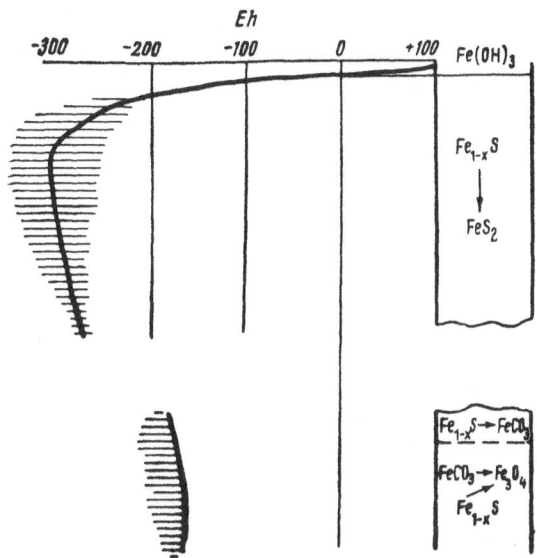

Fig. 59. Schematic picture of the change in oxidation—reduction potential during the formation of a sedimentary rock in which the following group of authigenic minerals was observed: $Fe_{1-x}S$ sulfides, pyrite, siderite, and magnetite.

and diagenesis may range rather widely. This fact raises doubt concerning electrometrical methods of measuring Eh in already-formed rocks for the purpose of determining the conditions under which they formed. Strictly, what is it that we may measure by the electrometrical method?

The difficulties of method arising in making such measurements and a critical examination of different methods for determining Eh are discussed rather fully by Sokolova [1962]. We shall not, therefore, analyze each method in detail. As Sokolova stated [Pustovalov and Sokolova, 1957; Sokolova, 1962], the most useful condition yielding stable results is measurement of Eh in samples moistened to a state of complete saturation of capillary capacity.

Soon after the work of Sokolova there appeared in print a detailed investigation by Solomin [1964], shedding doubt on the method of determining Eh based on saturation of a sample with water up to full capillary capacity. The doubt is associated with the possible charging of the sample by atmospheric oxygen. Solomin suggested special devices for measuring Eh, permitting one to avoid accidental inclusion of oxygen in the sample. In these devices, the samples were pulverized, placed in a current of nitrogen, and then, for better contact with the electrodes, were saturated with water to the state of a slurry. Strong arguments for the entire sequence of

processes lead us to believe the measurement of Eh by Solomin's method will be satisfactorily reliable. The only uncertainty now is the extent to which precisely obtained values of Eh reflect the actual oxidation—reduction state of the rock, especially for the period when the rock formed.

It is a fact that all methods proposed for direct determination of the oxidation—reduction potential in sedimentary rocks stop at the stage of achieving comparable results and make the a priori assumption that if comparable results are obtained with several measurements on a single sample this attests that the Eh obtained during the measurements corresponds to the oxidation—reduction state of the rock. One may readily convince himself of the incorrectness of this view.

Comparable results of several measurements on a single sample have been obtained in using several methods: Itkina [1952], Sokolova [1962], and Solomin [1964]. Nevertheless, each succeeding investigator finds cause for doubting the method of the preceding investigator. What guarantee do we have that the last method of Solomin makes it possible to set the sign of equality between the Eh measured and that which existed when the rock was formed? In the final analysis this also comes down to obtaining comparable results. It should be noted that in finding a theoretical basis for each method an important aspect escaped notice, a factor affecting the oxidation—reduction state of the rock. We have tried to emphasize this factor throughout this work: the effect of the composition and amount of gas on the oxidation—reduction potential. Any rock, if it has not been changed by metamorphism,[†] is characterized by definite contents of carbon dioxide and hydrogen sulfide, terminal relative to diagenesis. The complex equilibrium of minerals observed in the rock is due to these components. In all methods an inert gas is passed through the rock, and these natural gaseous components of the rock are removed or their relations are altered. As a result, the existing equilibrium is disturbed. This disturbance takes place rather easily because, as noted by Solomin [1964], the poise of the oxidation—reduction system in already-formed rock is extremely low. Therefore, a small change in equilibrium may lead to a great jump in potential because of the transition to a new potential-determining system. In support of this view we cite the results of experiments carried out by Sokolova [1962]:

[†]And, if it has been changed, in general there is then no sense in determining the oxidation—reduction potential.

"Several samples of rocks and ores were dried in air to the air-dried state, ground into powder, passed through a sieve with 0.5-mm openings, and then artificially moistened to the state of natural moisture content. The results of measuring Eh and pH in these samples showed that the oxidation-reduction potential in the artificially moistened samples was lower than that measured for the same samples but with natural moisture ... (p. 27). Eh and pH were measured in a strongly pyritized siliceous argillaceous rock twice: in its natural occurrence and after drying to the air-dried state and later saturation with water to a moisture content equal to full capillary capacity. The results of the two measurements were different. Very moist samples of siliceous argillite—phyllite rock, freshly taken from a borehole, gave off the odor of hydrogen sulfide and possessed marked reducing properties with Eh values from −290 to −395 mV at a pH from 6.3 to 8.0. When the Eh and pH of these samples were measured again, after drying and again moistening to full capillary capacity, i.e., after disturbing their natural gas and water conditions, the physicochemical properties of this rock had also changed: the potential had risen to 475 mV, but the pH had declined sharply" [Sokolova, 1962, p. 29] (emphasis mine, M. S.).

From these examples it was concluded that "irreversible processes occur during drying of the rock and remoistening, i.e., some of the components found in a dissolved form in the natural state did not return to solution after drying" [Sokolova, 1962, p. 27]. The above remarks restrict the use of electrometrical methods of determining the oxidation-reduction potential to recent muds and those rocks that possess sufficient moisture, permitting them to be measured without preliminary treatment. For other rocks and samples, supplementary indirect data are required.

Thus, in approaching the results of different methods of determining oxidation-reduction conditions under which sedimentary rocks formed, one must keep in mind that the greatest possibility for reconstructing such conditions is found in minerals with variable valence in their constituents. However, as we have tried to show here, the possibilities of these minerals are not reflected by the views recently developed in the works of Strakhov, Teodorovich, Garrels, Krumbein, or others. Mineral indicators of iron cannot of themselves give a unique solution even when the pH of the deposit has been determined by some features. For more precise limitation of the physicochemical conditions, apart from iron minerals, it is necessary to use some

other minerals with variable valence. In particular, if manganese minerals are present along with iron minerals, then, after investigating the nature of these minerals and plotting thermodynamic diagrams for them, we may obtain supplementary information, permitting us to draw more concrete boundaries for the medium relative to the oxidation-reduction conditions.

Without going into the details of an example, let us consider the manganese system and compare it with the iron system. A view of this has been thoroughly summarized for manganese oxides by Rode [1952]. The computations proposed here represent one of the possible, but not necessarily valid, variants, and its formal application without geologic and chemical verification may pass "into scientific speculation" as Krauskopf [1963] has expressed it.

It is first necessary to examine the question of the forms of manganese that will be found in solution, on the basis of which we determine the boundary of the solid phases. By using the summary of Pourbaix [1963], we may write down the following series of ions participating in the formation of solid compounds, and may also write the boundaries for each when the ionic ratio is equal to unity:

$$Mn^{..}/HMnO_2' \qquad pH = 11.46 \qquad (58)$$
$$Mn^{..}/Mn^{...} \qquad Eh = 1.509 \qquad (59)$$

$$
\begin{aligned}
Mn^{..}/MnO_4' && Eh &= 1.742 - 0.1182pH \\
HMnO_2'/MnO_4' && Eh &= 1.234 - 0.0738pH \\
Mn^{..}/MnO_4' && Eh &= 1.507 - 0.0945pH \\
Mn^{...}/MnO_4' && Eh &= 1.506 - 0.1182pH \\
MnO_4''/MnO_4' && Eh &= 0.564
\end{aligned}
$$

Concerning reaction (58), we should note that $a_{Mn^{..}}$ ions will increase with acidity of the solution. On the other hand, when alkalinity increases, $a_{HMnO_2'}$ ions increase. When $pH = 11.46$ $a_{Mn^{..}} = a_{HMnO_2'}$. With decline of pH one unit, the activity of $Mn^{..}$ increases one order relative to the activity of $HMnO_2'$. Therefore, toward alkaline solutions the boundary will be drawn relative to $a_{HMnO_2'}$, but toward acidic solutions, relative to $a_{Mn^{..}}$. As the relation in (59) shows, regardless of the pH value, both $Mn^{..}$ and $Mn^{...}$ ions may exist in the solution. Equal contents of the two occur at $Eh = 1.509V$. Since the complete equation has the form

$$Eh = 1.509 + 0.059 \lg \frac{a_{Mn^{...}}}{a_{Mn^{..}}}$$

with increase in Eh of 0.059 V, $a_{Mn^{..}}$ increases one order as compared with $a_{Mn^{...}}$. The upper boundary of water stability is determined by the equation

$$Eh = 1.229 - 0.059 pH$$

Setting these equations equal to each other and making simple transformations, we find

$$\frac{a_{Mn^{..}}}{a_{Mn^{...}}} = 10^{4.75 + pH}$$

i.e., even at the upper boundary of water stability and at pH = 0, $a_{Mn^{..}}$ is greater than $a_{Mn^{...}}$ by almost five orders.

With decline in Eh and increase in pH, this difference increases at a tremendous rate. Therefore, when plotting diagrams it is necessary to use $a_{Mn^{..}}$, not $a_{Mn^{...}}$. In similar fashion, in analyzing all the other reactions we may conclude that at the boundaries of water stability and in the fields of pH considered in this work, it is necessary to consider only $a_{Mn^{..}}$ for acidic conditions and only $a_{Mn^{...}}$ for alkaline conditions. On this basis, we shall furnish a series of equations, without deriving them, necessary for plotting stability fields of manganese oxides:

ACIDIC CONDITIONS

$$Mn^{..} + H_2O = MnO_2 + 2H^{.}, \qquad lg\ a_{Mn^{..}} = 15.31 - 2pH$$

$$3Mn^{..} + 4H_2O = Mn_3O_4 + 8H^{.} + 2e$$
$$Eh = 1.824 - 0.2364 pH - 0.0886\ lg\ a_{Mn^{..}}$$

$$2Mn^{..} + 3H_2O = Mn_2O_3 + 6H^{.} + 2e$$
$$Eh = 1.443 - 0.1773 pH - 0.0501\ lg\ a_{Mn^{..}}$$

$$Mn^{..} + 2H_2O = MnO_2 + 4H^{.} + 2e$$
$$Eh = 1.228 - 0.1182 pH - 0.0295\ lg\ a_{Mn^{..}}$$

EXISTENCE OF SOLID PHASES

$$3MnO + H_2O = Mn_3O_4 + 2H^{.} + 2e, \qquad Eh = 0.462 - 0.059 pH$$
$$2Mn_3O_4 + H_2O = 3Mn_2O_3 + 2H^{.} + 2e, \qquad Eh = 0.689 - 0.059 pH$$
$$Mn_2O_3 + H_2O = 2MnO_2 + 2H^{.} + 2e, \qquad Eh = 1.014 - 0.059 pH$$

AKALINE CONDITIONS

$$MnO + H_2O = HMnO_2' + H^{.}, \qquad lg\ a_{HMnO_2'} = -19.08 + pH$$

$$3HMnO_2' + H^{.} = Mn_3O_4 + 2H_2O + 2e$$
$$Eh = 1.228 + 0.059 pH - 0.0886\ lg\ a_{HMnO_2'}$$

$$2HMnO_2' = Mn_2O_3 + H_2O + 2e,\ Eh = -0.590 - 0.059\ lg\ a_{HMnO_2'}$$

Let us consider in comparison with this set of equations the conditions for formation of alabandite (MnS) when hydrogen sulfide represents the potential-determining system. For acidic conditions

$$MnS = Mn^{..} + S + 2e, \qquad Eh = -0.02 + 0.029\ lg\ a_{Mn^{..}}$$

The equation corresponding to the given reaction shows that when the activity of $Mn^{..}$ is less than 10 moles/liter the formation of alabandite is impossible, since the equilibrium line will lie below the boundary of water stability. Setting in parallel the formation of manganese oxides and alabandite gives the following results:

$$MnS + 2H_2O = MnO_2 + 4H^{.} + S + 4e$$
$$Eh = 0.60 - 0.059 pH$$

The transition $MnO_2 \rightarrow Mn_2O_3$ is effected at Eh = 1.014 - 0.059 pH, and, consequently, it begins under more strongly oxidizing conditions than alabandite begins to form. Relative to Mn_2O_3 the formation of alabandite may begin under the conditions

$$2MnS + 3H_2O = Mn_2O_3 + 6H^{.} + 2S + 6e$$
$$Eh = 0.46 - 0.059 pH$$

But Mn_2O_3 at Eh = 0.689 - 0.059 pH begins to go to Mn_3O_4, and, consequently, the possibility of alabandite forming here also fails to realize. It also fails to realize in all the remaining cases, as seen from this type of equation, since

$$3MnS + 4H_2O = Mn_3O_4 + 3S + 8H^{.} + 8e$$
$$Eh = 0.44 - 0.059 pH$$

$$MnS + H_2O = MnO + S + 2H^{.} + 2e$$
$$Eh = -0.02 - 0.059 pH$$

The last case determines the transition below the zone of water stability. Theoretically we may recognize the possible formation of alabandite in a strongly alkaline environment:

$$MnS + 2H_2O = HMnO_2' + S + 3H^{.} + 2e$$
$$Eh = 0.993 - 0.0885 pH + 0.0295\ lg\ a_{HMnO_2'}$$

However, this goes beyond the pH boundary considered in the present work. We must thus come to the conclusion that the MnS type of sulfide cannot form in normal sedimentary rocks. This conclusion is in agreement with actually observed facts, but does not correspond to theoretical computations proposed by Garrels [1960] or Krauskopf [1957] (see Figs. 10 and 11) for obvious reasons: these investigators used the irreversible reaction of hydrogen-sulfide oxidation to sulfates for their constructions.

Lastly, let us consider the formation of manganese carbonates. For this we may suggest the following series of reactions and their corresponding equations:

$$Mn^{..} + CO_2 + H_2O = MnCO_3 + 2H^{.}$$

$$2pH + \lg a_{Mn^{..}} + \lg P_{CO_2} = 8.1012$$

$$MnCO_3 + H_2O = MnO_2 + CO_2 + 2H^{.} + 2e$$
$$Eh = 0.989 - 0.059pH + 0.029 \lg P_{CO_2}$$

$$2MnCO_3 + H_2O = Mn_2O_3 + 2CO_2 + 2H^{.} + 2e$$
$$Eh = 0.964 - 0.059pH + 0.059 \lg P_{CO_2}$$

$$3MnCO_3 + H_2O = Mn_3O_4 + 3CO_2 + 2H^{.} + 2e$$
$$Eh = 1.102 - 0.059pH + 0.0885 \lg P_{CO_2}$$

$$MnO + CO_2 = MnCO_3, \qquad P_{CO_2} = 6.2 \cdot 10^{-8} \ atm$$

The diagrams shown in Figs. 60 and 61 were plotted by means of these equations and also those calculations for iron minerals given in this work for the case when hydrogen sulfide represents the potential-determining system. Among other things, we may clearly see from Figs. 60 and 61 that manganese compounds are much more soluble than iron compounds, and, consequently, manganese ions have

a much greater capacity for migration at any value of Eh. With the joint occurrence of two syngenetic or diagenetic minerals, such diagrams permit a more detailed classification of the medium in which the sediments accumulated and consolidated according to Eh, pH, P_{CO_2}, and other parameters.

Let us make a parallel analysis of syngenetic and diagenetic formation of minerals, the specific aspects of which we have already considered relative to Eh and pH, and the formation of minerals in the supergene zone. Such a comparison is now possible because of the work of Perel'man [1965], who generalized the results of geologic and mineralogical observations.

It has been shown how, during mineral growth in bottom deposits as well as in the supergene zone, the specific characteristics of the minerals depend in great part on the effect of the gaseous components of the solutions: oxygen, carbon dioxide, and hydrogen sulfide. This effect is sufficiently appreciable, to the extent that it seems appropriate to distinguish three types of environments in the supergene zone also: 1) oxidizing, 2) reducing without hydrogen sulfide (gleyey in the terminology of Perel'man), and 3) reducing sulfide. As Perel'man has shown [1965], "the difference between the latter two environments is determined by the content of hydrogen sulfide, not by the value of the oxidation–reduction potential. At any specific low potential, according to the H$_2$S content, we may find

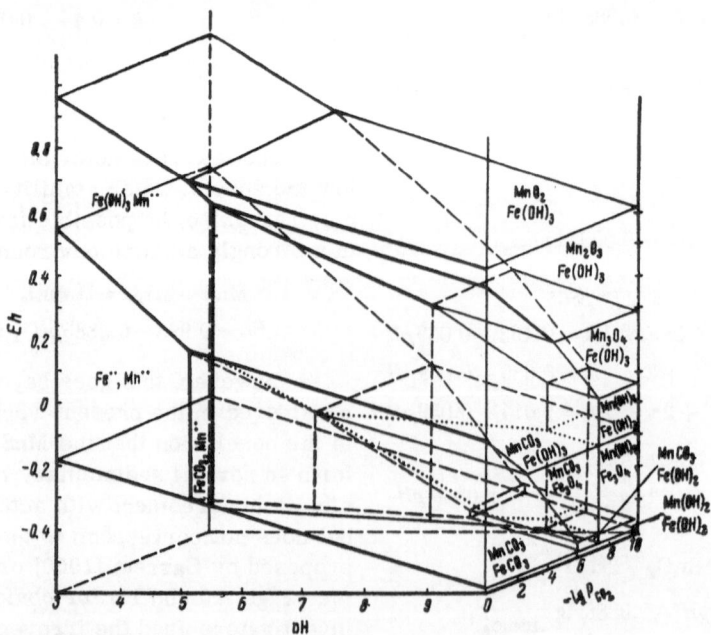

Fig. 60. Paragenesis of iron and manganese minerals at t = 25°C, P = 1 atm, $a_{\sum Fe} = a_{\sum Mn} = 10^{-4}$ mole/liter. Reduction of the sulfate ion is lacking.

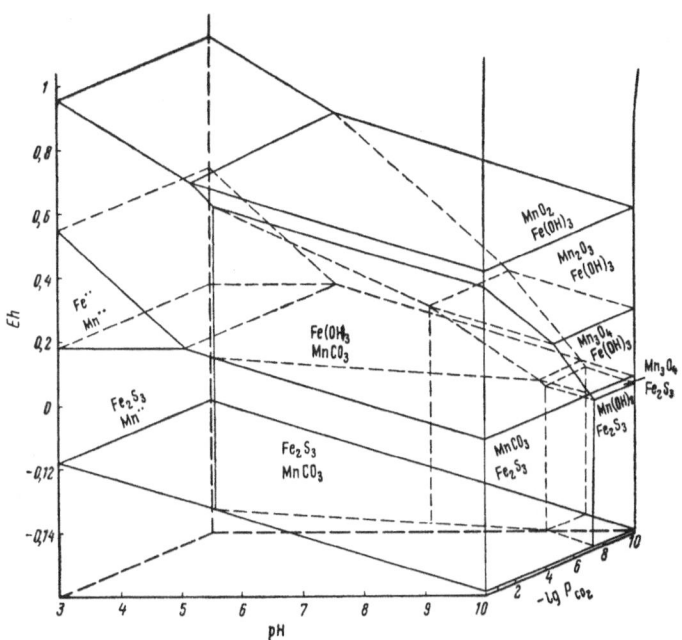

Fig. 61. Paragenesis of iron and manganese minerals at t = 25°C, $a_{\Sigma Fe} = a_{\Sigma Mn} = 10^{-3}$ mole/liter. Hydrogen sulfide potential-determining system.

the second environment (such as on swampy taigas) or the third (such as in salt-pan deserts)."

Thus, despite the differences in geologic relations of the environment, geochemical and micro-biological factors remain the same, determining a single specific group of conditions for the formation of minerals in the sedimentary shell of the earth.

CONCLUSIONS

We have tried to evaluate the principal methods of determining the oxidation–reduction conditions under which sedimentary rocks form. In analyzing each of these methods, it has been concluded that not one of them gives reliable information concerning the exact oxidation–reduction potential for the medium during accumulation and diagenesis of sediments. The most usable method, permitting us to reach definite conclusions concerning oxidation–reduction conditions during the formation of a sedimentary rock, is detailed petrographic study of the rocks with determination of the sequence in which the minerals formed. Unfortunately, the petrographic method, the most important in geologic investigations, has been used less and less in geologic investigations in recent years, giving way to bulk comparisons of organic matter, the ratios of the forms of iron, sulfur, and other elements.

In the study of iron minerals, it is possible generally to investigate only one boundary for Eh, i.e., to determine whether the oxidation–reduction potential was higher or lower than some particular value. For narrowing the boundaries it is necessary to investigate the sequence not only of iron minerals but of other minerals also, minerals that contain elements with variable valence. Iron minerals are the most widespread; they are almost universal.

The author reports that not all statements put forth in this work have equally firm foundations. Many solutions do not go beyond the framework of conjecture. But the book contains some conclusions that, in the author's opinion, are beyond doubt.

1. The use of thermodynamic constructions with the control of geologic facts may give a precise (and not approximate) picture of the formation of minerals.

2. The gaseous components of the system have a strong effect on the processes of mineral formation. It is therefore necessary to account for the effect of these components for specific oxidation–reduction conditions during the formation of rocks. They define two types of mineral formation: gleyey and sulfide.

3. In the gleyey type of mineral formation, siderite and magnetite may change places in the reduction series for reproducibility of conditions, depending not so much on Eh as P_{CO_2}.

4. In the sulfide type of mineral formation, hydrogen sulfide represents the potential-determining system for the period of syngenesis and early diagenesis. The sulfide ion that forms does not conform to the thermodynamic aspects of the active part of the medium. This leaves its impression on the sequence of mineral formation, leading to the result that iron sulfide (pyrite) will appear as the first mineral after iron hydroxides in the reduction series when the oxidation–reduction potential declines.

5. During diagenesis, the oxidation–reduction potential is found to rise and not decline, accompanied by an uncompensated decrease of the sulfide ion. Therefore, in the sulfide type of mineral formation, it is possible for diagenetic siderite and magnetite to appear.

6. The most acceptable method for determining the oxidation–reduction conditions during formation of sedimentary rocks is at present the method of studying the sequence of mineral formations for at least two different types of minerals (such as manganese and iron minerals).

The above conclusions are the basis for making the following statements.

1. The ratio of the forms of iron or sulfur or the quantitative evaluation of organic matter cannot be considered sufficiently reliable criteria of oxidation–reduction conditions for the formation of rocks.

2. The geochemical series of minerals, proposed earlier by Pustovalov as a working scheme, reflecting the sequence of changes in minerals according to Eh values, played a positive role, but at this time cannot be used, since, in constructing the series, special features in the behavior of the oxi-

dation–reduction potential during diagenesis and the effect of gaseous components of the medium were not accounted for.

3. The diagram proposed by Krumbein and Garrels, being based only on formal thermodynamic calculations and not being corrected by geologic aspects of the environment, does not reflect the sequential change in mineral formation during the period of sedimentation or diagenesis.

4. Electrometrical methods are most convenient to use for investigation of secondary oxidation–reduction processes that take place in rocks because of hydrogeologic factors. Mostly (except for recent sediments) the electrometrical methods do not reflect the oxidation–reduction conditions of sedimentation and diagenesis.

For a successful solution of the problem of oxidation–reduction potential in geology, it is first necessary to conduct work to refine the thermodynamic constants of minerals. A great aid in this might be rendered by investigations on mineral syn-

thesis. There is special importance in refining our knowledge of the conditions under which different types of sulfides form and in determining the possible range of x in Fe_{1-x} Se type sulfides. In this connection, it seems to us advisable to set up special programs for studying phase analysis of sulfides, and for investigating the possibilities of hydrogen sulfide as a potential-determining system. Such programs lead to well-rounded and well-equipped approaches to the determination of the physico-chemical conditions under which sedimentary rocks form and, ultimately, to the prediction of mineral resources.

If the author of the present book has succeeded, even partially, in convincing the reader of the validity of the ideas put forward, he will consider the time spent in preparing this book not to have been in vain. If the reader retains his own views, the author will accept with gratitude all critical remarks and objections, which will be taken into account in his daily scientific work.

APPENDIX

Values of ΔG for Some Common Compounds and Ions

Formula	State	ΔG^o, kcal	Source
H_2O	Liquid	-56.69	Latimer [1952]
OH'	Aqueous	-57.595	"
$H^{\cdot}$	"	0	»
$Fe^{\cdot\cdot}$	»	-20.30	
$Fe^{\cdot\cdot\cdot}$	»	-3.00	Computed data
Fe_3O_4	Amorphous	-232.04	" "
$Fe(OH)_2$	Crystalline	-115.57	Latimer [1952]
$Fe(OH)_3$	»	-169.45	Computed data
FeS_α	»	-23.32	Latimer
$Fe_{1-x}S$	»	$\dfrac{(23.32 + 39.84\,x)}{(1 + 2\,x)}$	Computed data
FeS_2 (pyrite)	»	-39.84	Latimer
$FeCO_3$	»	-161.06	»
S (orthorhombic)	»	0	»
S_2	Gas	19.96	Computed data
S''	Aqueous	21.958	Pourbaix [1963]
S_2''	»	19.749	»
SO_4''	»	-177.34	Latimer
HS'	»	3.01	»
H_2S	Gas	-7.892	»
H_2S	Aqueous	-6.54	»
H_2CO_3	»	-149.00	»
HCO_3'	»	-140.31	»
CO_3''	»	-126.22	»
CO_2	»	-92.31	»
CO_2	Gas	-94.2598	»

111

LITERATURE CITED

Aleshina, V. I., "Destruction of chitin by sulfate-reducing bacteria and changes in the oxidation-reduction conditions during sulfate reduction," Mikrobiologiya, Vol. 7, No. 7 (1938).

Aleskina, I. A., "Mineralogy of the coarse-silt fraction of bottom sediments from Kronoki and Avacha Gulfs," Tr. Inst. Okeanologii AN SSSR, Izd. AN SSSR, Vol. 61 (1962).

Andrusov, N., "Some results of the 'Chernomortsa' expedition. On the origin of hydrogen sulfide in waters of the Black Sea," Izv. Imp. Russk. Georg. Obshch., Vol. 28 (1892).

Angel, F., and Scharizer, R., Grundriss der Mineralparagenese, Wien (1952).

Arkhangel'skii, A. D., "Iron sulfide in deposits of the Black Sea," Byull. MOIP, Otd. Geol., Vol. 12, No. 3 (1934).

Baas Becking, L. G. M., "Biological processes in the estuarine environment. VI. The state of the iron in the estaurine mud iron sulfides," Proc. Koninkl. Nederl. Akad. Wet., Vol. 3 (1956).

Baas Becking, L. G. M., Kaplan, I. R., and Moore, D., "Limits of the natural environment in terms of pH and oxidation-reduction potentials," Jour. Geol., 68:243 (1960).

Baas Becking, L. G. M., and Moore, D., "Biogenic sulfides," Econ. Geol., Vol. 56, No. 2 (1961).

Babaev, A. G., "The geochemical range in the formation of pyrite and glauconite in sedimentary rocks," Dokl. AN ArmSSR, Vol. 25, No. 3 (1957a).

Babaev, A. G., "Paragenesis and generations of authigenic minerals (on the basis of Cretaceous rocks in western Uzbekistan)," Izv. AN ArmSSR, Ser. Geol. i Geogr. Nauk, Vol. 10, No. 5-6 (1957b).

Bardoshi, D., and Bod, M., "A new method of measuring oxidation-reduction properties of sedimentary rocks," Geokhimiya, No. 3 (1960).

Barker, H. A., "On the biochemistry of methane fermentation," Archiv. Mikrobiol., Vol. 7, No. 4 (1936).

Barnes, H. L., and Kullerud, G., "Equilibria in sulfur-containing aqueous solutions in the system Fe-S-O and their correlation during ore deposition," Econ. Geol., Vol. 56, No. 4 (1961).

Behrend, F., and Berg, G., Chemische Geologie, Stuttgart (1927).

Beijerinck, W. M., "Über Spirillum desulfuricans als Ursache von sulfat--reduktion," Zbl. fur Bakteriol., Pt. II, Vol. 1 (1895).

Berner, R. A., "Electrode studies of hydrogen sulfide in marine sediments," Geochim. et Cosmochim. Acta, Vol. 27, No. 6 (1963).

Berner, R. A., "Distribution and diagenesis of sulfur in some sediments from the Gulf of California," Marine Geol., Vol. 1, No. 1 (1964a).

Berner, R. A., "Iron sulfides formed from aqueous solution at low temperatures and atmospheric pressure," Jour. Geol., Vol. 72, No. 3 (1964b).

Berner, R. A., "Stability fields of iron minerals in anaerobic marine sediments," Jour. Geol., Vol. 72, No. 6 (1964c).

Berz, K. S., "Über Magnetkisen in Marinenablagerungen," Zbl. Miner., Vol. 18 (1922).

Bobrovnik, D. P., Barite and Pyrrhotite in Tortonian Rocks on the Southwestern Margin of the Russian Platform. Questions on the Mineralogy of Sedimentary Rocks, Book 5, Izd. L'vovskogo Univ. (1958).

Bogdanova, A. K., Water Circulation through the Bosporus and Its Role in Mixing the Waters of the Black Sea, Tr. Sevastopol'skoi Biol. Stantsii, Vol. 12, Sevastopol' (1959).

Bogdanova, A. K., New Data on the Distribution of Mediterranean Waters in the Black Sea, Tr. Sevastopol'skoi Biol. Stantsii, Vol. 13, Izd. AN SSSR (1960).

Bogdanova, A. K., "Distribution of Mediterranean Waters in the Black Sea," Okeanologiya, Vol. 1, No. 6 (1961).

Böggild, O. B., On the Bottom Deposits of the North Polar Sea. Scientific Results of the Norwegian North Polar Expedition, London (1906).

Bonshtedt-Kupletskaya, É. M., "Activities of the Committee on New Minerals and Mineral Names," International Mineralogical Association (1967–69), Zap. Vsesoyuzn. Mineralogich. Obshch., Ser. 2, Part 99, No. 4 (1970).

Braconnt, H., "Examen de la boue noire provenant des égouts," Ann. Chim. et Phys., Vol. 50 (1832).

Bruevich, S. V., Chemical Investigations of the Oceanological Institute (IOAN) in Far Eastern Seas and Adjoining Parts of the Pacific Ocean, Tr. Inst. Okeanologii AN SSSR, Vol. 17, Izd. AN SSSR (1956).

Bruevich, S. V., and Zaitseva, E. D., "Chemistry of the sediments in the Bering Sea," Tr. Inst. Okeanologii AN SSSR, Vol. 31, Izd. AN SSSR (1958).

Bushinskii, G. I., "Petrography and some questions on the origin of the Vyatka phosphorites," Byull. MOIP, Otd. Geol., Vol. 14 (2) (1936).

Bushinskii, G. I., Geochemistry of the Sedimentary Process. A Field Guide for Petroleum Geologists, Vol. 1, Gostoptekhizdat (1954).

Chukhrov, F. V., "Pyrrhotite and pyrite in the Kerch ores and some general questions on the origin of iron sulfides," Izd. Akad. Nauk SSSR, Ser. Geol., No. 1 (1936).

Chukhrov, F. V., Genkin, A. D., Soboleva, S. V., and Basov, G. V., "Smythite from iron-ore deposits of the Kerch Peninsula," Litolog. i Polezn. Iskop., No. 2 (1965).

Dal'nichenko, P. T., and Chigirin, N. I., "The origin of hydrogen sulfide in the Black Sea," Tr. Osoboi Zool. Lab. i Sevast. Biol. Stantsii AN SSSR, Ser. 2, No. 10, Izd. AN SSSR (1926).

Debyser, J., "Relation entre le pH, le Eh, et la diagenese," Rev. Inst. Franc. Pétrole, Vol. 12 (1957).

Doelter, C., "Hydrotroilit," in: C. Doelter and H. Leitmeier, Handbuch d. Mineralchemie (1926).

Doss, B. O., "Three gas wells on the property of the Mel'nikov brothers in the Samara Province," Ezhegodnik po Geol. i Mineralogii Rossii, Vol. 12, No. 5–6 (1911).

Durov, S. A., Purification of Drinking Water from Hydrogen Sulfide, Rostov-na-Donu (1935).

Durov, S. A., and Turzheva, M. P., Determination of Total Hydrogen-Sulfide Content and Dissolved Oxygen in Water, Gidrokhim. Mat., Vol. 13, Izd. AN SSSR (1947).

Egunov, M. A., "Bio-anisotropic basins," Ezhegodnik po Geol. i Mineralogii Rossii, No. 4, Novaya Aleksandriya (1900–01).

Ehrenberg, H., Neues Jahrb. Mineral Geol. und Paläontol., Vol. 57, Pt. 2, Stuttgart (1928).

Ékzertsev, V. A., "Determination of the thickness of the microbiologically active layer of muds in several lakes," Mikrobiologiya, Vol. 17, No. 6 (1948).

Emery, K. O., and Rittenberg, S. C., "Early diagenesis of California basin sediments in relation to the origin of oil," Bull. Am. Assoc. Petrol. Geologists, Vol. 36, No. 5 (1952).

Epatko, Yu. M., and Shnyukov, E. F., Conditions of Formation of Carbonate Concretions in the Kerch Basin, Zap. Ukr. Otd. Vsesoyuz. Min. Obshch., Izd. AN UkrSSR, Kiev (1962).

Erd, R. C., and Evans, H. T., "The compound Fe_3S_4 (smythite) found in nature," J. Amer. Chem. Soc., Vol. 18 (1956).

Erd, R. C., Evans, H. T., and Richter, D. N., "Smythite, a new iron sulfide, and associated pyrrhotite from Indiana," Amer. Mineralogist, Vol. 42, Nos. 5–6 (1957).

Evseeva, L. S., and Fomina, N. P., Oxidation–Reduction Properties of Sedimentary Uraniferous Rocks, Atomizdat (1965).

Evseeva, L. S., and Perel'man, A. I., Geochemistry of Uranium in the Supergene Zone, Atomizdat (1962).

Fersman, A. E., Selected Works, Vol. 3, Izd. AN SSSR.

Galliher, E. W., "Geology of glauconite," Bull. Am. Assoc. Petrol. Geologists, Vol. 19, No. 11 (1935).

Garrels, R. M., Mineral Equilibria, Harper and Row, New York (1960).

Garrels, R. M., and Christ, C. L., Solutions, Minerals, and Equilibria, Harper and Row, New York (1965).

Ginzburg, I. I., and Rukavishnikova, I. A., Minerals of Ancient Weathering Zones in the Urals, Izd. AN SSSR (1951).

Ginzburg-Karagicheva, T. L., The Role of Bacteria in the Processes of Oil Formation, VNIGNI, No. 4, Gostoptekhizdat (1954).

Godlevskii, M. N., A Method of Constructing Physicochemical Diagrams, Izd. Nedra (1965).

Gol'dshmidt (Goldschmidt), V. M., "History of the iron-family minerals in nature," in: Basic Ideas in Geochemistry, Goskhimtekhizdat (1933).

Gololobov, Ya. K., Thickness of the Oxygen–Hydrogen Sulfide Layer in the Black Sea, Gidrokhim. Materialy, Vol. 21, Izd. AN SSSR (1953).

Golovin, F. I., Thermodynamic Constant of the First Stage of Hydrogen-Sulfide Dissociation in Water at Various Temperatures, Gidrokhim. Materialy, Vol. 29, Izd. AN SSSR (1959).

Gordeev, E. I., "Determination of hydrogen-ion concentrations and the oxidation–reduction potential in marine sediments," Tr. Inst. Okeanologii, Vol. 4, Izd. AN SSSR (1962).

Gorshkova, T. I., "Chemical–mineralogical investigation of sediments in the Barents and White Seas," Tr. Gos. Okeanograf. Inst., Vol. 1, No. 2–3, Gostekhizdat (1931).

Gulyaeva, L. A., "Geochemical facies, oxidation–reduction conditions, and organic matter in sedimentary rocks," Sovetskaya Geologiya, No. 47 (1955).

Gulyaeva, L. A., Geochemistry of Devonian and Carboniferous Rocks in the Kuybyshev Segment of the Volga District, Izd. AN SSSR (1956).

Huber, N. K., and Garrels, R. M., "Relations of pH and oxidation potential to sedimentary iron mineral formation," Econ. Geol., Vol. 48, p. 337 (1953).

Isachenko, B. L., Investigations on the Bacteria of the Arctic Ocean, Tr. Murmansk. Nauch. Promyslovoi Expeditsii 1906 g., Petrograd (1914).

Isachenko, B. L., On the Origin of Sulfur, Tr. Inst. Mikrobiologii AN SSSR, No. 5, Izd. AN SSSR (1958).

Isachenko, B. L., and Egorova, A., "The bacterial layer in the Black Sea," in: Papers in Honor of N. M. Knipovich, Pishchepromizdat (1939).

Isaeva, A. B., "Geochemical investigations of sediments in the northern part of the Indian Ocean," Tr. Inst. Okeanologii AN SSSR, Vol. 64, Izd. AN SSSR (1964).

Ismailov, M. A., "Lithology of the Chokrak Deposits of Southern Kobystan (Kyrkishlak region)," Azerbaidzh. Neft. Khozyaist-vo, No. 10 (400) (1959).

Itkina, E. S., Methods of Determining the Oxidation–Reduction Potential in Rocks, Tr. Inst. Nefti AN SSSR, Vol. 2, Izd. AN SSSR (1952).

Ivanov, K. B., "Authigenic pyrrhotite from Lower Carboniferous Rocks in the Vicinity of Tomsk," Dokl. UP. Nauch. Konf. Posvyashch. 40-letiyu Velikoi Oktyabr'skoi Revol., No. 4 (1957).

Ivanov, M. V., "Use of isotopes in studying intensity of the sulfate-reducing process in Lake Belovod'," Mikrobiologiya, Vol. 25, No. 3 (1956).

Ivanov, M. V., "The role of microorganisms in the formation of sulfur deposits in the hydrogen-sulfide springs of the Sergievsk mineral waters," Mikrobiologiya, Vol. 26, No. 3 (1957).

Ivanov, M. V., Study of the Intensity of the Sulfur Cycle in Lakes by Means of Radioactive Sulfur (S^{35}), Tr. VI Soveshch. po Problemam Biol. Vnutrennikh Vod., Izd. AN SSSR (1959).

Ivanov, M. V., "Microbiological investigations of the Carpathian sulfur deposits. Study of the microbiological process of sulfate reduction in the Rozdol sulfur deposit," Mikrobiologiya, Vol. 24, No. 2 (1960).

Ivanov, M. V., The Role of Microorganisms in the Formation and Destruction of Sulfur Deposits, Tr. Inst. Mikrobiologii AN SSSR, No. 9, Izd. AN SSSR (1961).

Ivanov, M. V., The Role of Microbiological Processes in the Origin of Deposits of Native Sulfur, Izd. AN SSSR (1964).

Ivanov, M. V., and Terebkova, L. S., "Study of the microbiological processes in the formation of hydrogen sulfide in a saline lake," Mikrobiologiya, Vol. 28, No. 2 (1959).

Kalganov, M. I., "The Maloe Khalilovo deposit," in: The Khalilovo Deposit of Complex Iron Ores, Tr. Inst. Geol. Nauk, No. 67, Izd. AN SSSR (1942).

Kanel', É. S., "The oxidation–reduction potential of a medium as the limiting factor in development of microorganisms," Mikrobiologiya, Vol. 6, No. 2 (1937).

Kaplan, I. R., Emery, K. O., and Rittenberg, S. C., "The distribution and isotopic abundance of sulfur in recent marine sediments off Southern California," Geochim. et Cosmochim. Acta, Vol. 27 (1963).

Karapet'yants, M. Kh., Methods of Comparative Computation of Physicochemical Properties, Izd. Nauka (1965).

Kaz'mina, T. I., Maimin, Z. L., and Petrova, Yu. N., "The determination by geochemical indicators of the conditions of formation of sediments in Devonian basins in the northwest part of the Russian platform," Tr. VNIGNI, No. 95, Gosgeoltekhizdat (1956).

Khalatin, N. V., "Mineralogy and geochemistry of the Lower Carboniferous terrigenous sequence in the southern part of the Kama-Kinel' basin," Sovetskaya Geologiya, No. 8 (1961).

Klenova, M. V., Instructions for Determining the Ratio of Ferric and Ferrous Oxides and for Extracting Soil Solution, Instruktsiya Gos. Okeanogr. Inst. Sektor Geol. Morya, No. 12, Gostekhteorizdat (1933).

Klenova, M. V., Sediments of the Motovskii Gulf (On the Question of Joint Investigations of Recent Marine Sediments), Tr. Vsesoyuz. Nauch.-issled. Inst., Morskogo Rybnogo Khozyaistva i Okeanogr., Vol. 5 ONTIZ NKTP SSSR (1938).

Klenova, M. V., Geology of the Sea, Uchpedgiz (1948).

Klenova, M. V., "Fundamental patterns of marine sedimentation," Izv. Akad. Nauk SSSR, Ser. Geol. (1951).

Klenova, M. V., "Processes of sedimentation on the underwater slope of the Caspian Sea in Azerbaidzhan," in: Recent Sediments of the Caspian Sea, Izd. AN SSSR (1956).

Klenova, M. V., "Sediments of the Arctic Basin from material collected during drift of the ice ship Georgii Sedov," Izv. AN SSSR (1962).

Kliburszky, B., "Die physikalischen Grundlagen der geochemischen Potenzialberechnung," Acta Geol. Acad. Scient. Hung., Vol. 5, No. 3–4 (1958).

Knipovich, É. V., "Iron ores," in: Analyses of Mineral Raw Materials, ONTI-khimteoret, Leningrad (1936).

Korolev, D. F., and Kozerenko, S. V., "Experimental study of the conditions for formation of iron sulfides from solution," Dokl. Akad. Nauk SSSR, Vol. 165, No. 6 (1965).

Kramer, V. A., and Vail', E. I., "Potentiometric determination of metals in sulfide form by using natural materials for reference electrodes," Zavodskaya Laboratoriya, Vol. 23, No. 2 (1957).

Krauskopf, K. B., "Separation of manganese from iron in sedimentary processes," Geochim. et Cosmochim., Vol. 12, 61 (1957).

Kriss, A. E., Marine Microbiology (Deep Water), Izd. AN SSSR (1959).

Krumbein, W. C., and Garrels, R. M., "Origin and classification of sediments in terms of pH and oxidation–reduction potentials," Jour. Geol., Vol. 60, 1 (1952).

Kryukov, P. A., The Oxidation–Reduction State of Water in the Group of Caucasian Mineral Waters, Gidrokhim. Materialy, Vol. 14, Izd. AN SSSR (1948).

Kryukov, P. A., Zavodnov, S. S., and Goremykin, V. É., "Sulfide-carbonate equilibrium and the oxidation–reduction state of sulfur in mineral waters of the region of the Caucasian mineral waters," Dokl. Akad. Nauk SSSR, Vol. 142, No. 1 (1962).

Kullerud, G., and Yoder, H. S., "Pyrite stability relations in the Fe–S system," Econ. Geol., Vol. 54, No. 4 (1959).

Kumai, T., "Investigation of the reaction of ferric hydroxide with hydrogen sulfide," Nihon Kagaku Zasshi, Vol. 78, No. 8 (1957).

Kumai, T., "Reaction between ferric hydroxide and hydrogen sulfide," Nihon Kagaku Zasshi, Vol. 79, No. 6 (1958).

Kuznetsov, M. D., and Sagalovskii, A. É., "Kinetics of the process of hydrogen-sulfide absorption by ferric hydroxide," Zh. Prikl. Khim., Vol. 27 (1954).

Kuznetsov, S. I., Ivanov, M. V., Lyalikova, N. N., Introduction to Geological Microbiology, Izd. AN SSSR (1962).

Kuznetsov, S. I., and Sokolova, G. A., "Some data on the physiology of Thiobacillus thioparus," Mikrobiologiya, Vol. 24, No. 2 (1960).

Kuznetsova, Z. I., Distribution and Ecology of Microorganisms in Deep Underground Waters in Some Parts of the USSR, Tr. Inst. Mikrobiologii, AN SSSR, No. 9, Izd. AN SSSR (1961).

Lapteva, O. N., "Dependence of the oxidation–reduction potential of a solution containing ferric and ferrous ions on pH," Zh. Prikl. Khim., Vol. 31, No. 8 (1958).

Laput', V. A., "Content of sulfur in Devonian rocks of the Pripyat' Ridge as an indicator of oxidation-reduction conditions of sedimentation," DAN BSSR, Vol. 6, No. 7 (1962).

Larskaya, E. S., "Geochemical facies in Mesozoic rocks of eastern Ciscaucasia," Dokl. Akad. Nauk SSSR, Vol. 140, No. 6 (1961).

Latimer, W. M., Oxidation Potentials, 2d ed., Prentice Hall, New York (1952).

Lebedintsev, A. A., Preliminary Account of Chemical Investigations of the Black Sea and the Sea of Azov in the Summer of 1891, Izv. Imp. Russk. Geogr. Obshch., Vol. 28 (1892).

Lebedintsev, A. A., Preliminary Account of the Chemical Investigations of the Black Sea and the Sea of Azov in the Summer of 1891, Zap. Novorossiiskogo Obshch. Estestvoispytatelei, Vol. 6 (1892).

Lees, H., Biochemistry of Autotrophic Bacteria, Butterworths, London (1955).

Lepp, H., "The synthesis and probable geologic significance of melnikovite," Econ. Geol., Vol. 52, No. 5 (1957).

Levchenko, V. M., The Oxidation–Reduction Potential of the Sulfur System, Gidrokhim. Materialy, Vol. 18, Izd. AN SSSR (1950).

Levchenko, V. M., and Makarova, K. A., The Oxidation of Sulfides, Tr. Khim. Inst. Kirgiz. Fil. AN SSSR, No. 3, Frunze (1950).

Levinson, V. É., Geochemical Bituminology and Its Problems, Vol. 4, Izd. Nauka (1964).

Lindren, W., "The colloid chemistry of minerals and ore deposits," in: Theory and Application of Colloidal Behavior, ed. by R. H. Bande, Vol. 2 (1926).

Litvinenko, A. U., and Drozdov, G. M., "Supergene magnetite from a weathering zone on ultrabasic rocks in the middle Dnieper region," Dokl. Akad. Nauk SSSR, Vol. 145, No. 2 (1962).

Lukashev, K. I., Outlines of the Geochemistry of Supergene Activity, Minsk (1963).

Maronny, G., and Valensi, G., "Fonctions thermodynamiques standart des ions mono et polisulfures en solution aqueuse," 9me Reunion du C.I.T.C.E., Paris (1957).

Mason, B., "Oxidation and reduction in geochemistry," J. Geol., Vol. 57, pp. 62–72 (1942).

Michaélis, L., Oxidation–Reduction Potentials, Lippencott, Philadelphia and London [1930].

Mishina, T. A., Avdeeva, O. I., and Bozhovskaya, T. K., "The solubility of gases in natural waters according to temperature, pressure, and salt composition," Materialy VSEGEI, Nov. Ser., No. 46, Gosgeoltekhizdat (1961).

Mislovitser, É. É., Determining the Concentration of Hydrogen Ions in Liquids, Goskhimtekhizdat (1930).

Mokievskaya, V. V., Some Hydrochemical Aspects of the Indian Ocean, Okeanologicheskie Issledovaniya, No. 4 (1961).

Nadeinskii, B. P., Theoretical Foundations and Computations in Analytical Chemistry, Izd. Sovetskaya Nauka (1956).

Ontoev, D. O., "Composition and conditions of formation of iron chlorites in some hydrothermal deposits," Izv. Akad. Nauk SSSR, Ser. Geol., No. 4 (1956).

Ostroumov, É. A., "Method of determining the forms of sulfur compounds in deposits of the Black Sea," Tr. Inst. Okeanologii, Vol. 7, Izd. AN SSSR (1953a).

Ostroumov, É. A., "The forms of sulfur compounds in deposits of the Black Sea," Tr. Inst. Okeanologii, Vol. 7, Izd. AN SSSR (1953b).

Ostroumov, É. A., "Sulfur compounds in the bottom deposits of the Sea of Okhotsk," Tr. Inst. Okeanologii, Vol. 12, Izd. AN SSSR (1957).

Ostroumov, É. A, and Fomina, L. S., "Forms of sulfur compounds in bottom deposits of the Marianas trench," Dokl. Akad. Nauk SSSR, Vol. 126, No. 2 (1959).

Ostroumov, É. A., and Volkov, I. I., "Forms of sulfur compounds in bottom deposits of the Pacific Ocean near New Zealand," Tr. Inst. Okeanologii, Vol. 63, Izd. AN SSSR (1960).

Ostroumov, É. A., Volkov, I. I., and Fomina, L. S., "Distribution of the forms of sulfur compounds in bottom deposits of the Black Sea," Tr. Inst. Okeanologii, Vol. 50, Izd. AN SSSR (1961).

Ovsyannikova, K. A., "Oxidation-reduction potential of the salt waters and soils of Lake Sak," Gidrokhim. Mat., Vol. 19, Izd. AN SSSR (1951).

Pavlov, A. L., "Chemistry of siderite formation in hydrothermal deposits," Geol. i Geofiz., No. 4 (1964a).

Pavlov, A. L., "Low-temperature hydrothermal synthesis of hematite and magnetite," Geol. i Geofiz., No. 11 (1964b).

Pel'sh, A. D., Energy of the Desulfuration Process, Tr. Solyanoi Labor. Vsesoyuzn. Inst. Galurgii, No. 14, Goskhimizdat (1937).

Perel'man, A. I., Geochemistry of Epigenetic Processes, Izd. Nedra (1965).

Poddubnyi, I. I., "Dynamics of oxidation—reduction conditions in saline soils of the Atkarsk region of the Saratov Oblast," Dokl. TSKhA, No. 42 (1959).

Polushkina, A. P., and Sidorenko, G. A., "Melnikovite as a mineral species," Zap. Vsesoyuz. Min. Obshch., Ser. 2, Pt. 92 (1963).

Polushkina, A. P., and Sidorenko, G. A., "Melnikovite should be considered a definite mineral spec ies," Zap. Vsesoyuzn. Mineralogich. Obshch., Pt. 97, No. 3 (1968).

Postgate, J. R., "The reduction of sulfur compounds by *Desulphovibrio desulphuricans*, J. Gen Microbiol., Vol. 5, No. 4 (1951).

Pourbaix, M., Atlas d'equilibres Electrochimiques, Paris (1963).

Pustovalov, L. V., "Basic features of the geochemical processes during the formation of sedimentary rocks," Tr. Vsesoyuz. Geol.-razved. Ob'edineniya NKTP SSSR, No. 285, Gosgeolizdat (1933a).

Pustovalov, L. V. "Geochemical facies and their significance in general and applied geology," Problemy Sovetskoi Geologii, Vol. 1, No. 1 (1933b).

Pustovalov, L. V., Petrography of Sedimentary Rocks, Gostoptekhizdat (1940).

Pustovalov, L. V., and Sokolova, E. I., "Methods of determining pH and Eh in sedimentary rocks," in: Methods of Studying Sedimentary Rocks, Vol. 2, Izd. AN SSSR (1957).

Rabotnova, I. L., The Role of Physicochemical Conditions (pH and rH_2) in the Life Activity of Microorganisms, Izd. AN SSSR (1957).

Rabotnova, I. L., Toropova, E. G., and Rabaeva, M. Yu., "Requirements of anaerobic bacteria relative to the oxidation—reduction conditions of the environment," Mikrobiologiya, Vol. 24, No. 5 (1955).

Rakhmanov, V. P., "Supergene magnetite in the weathering crust of ferruginous quartzites of the Kursk Magnetic Anomaly," Dokl. Akad. Nauk SSSR, Vol. 122, No. 6 (1958).

Ramdohr, R., Die Erzmineralien und ihre Verwachsungen, Akademie Verlag, Berlin (1950—60).

Rauzer-Chernousova, D. M., "Geologic investigation of a salt lake at Krugloe Bay near Sevastopol'," Izv. Akad. Nauk SSSR, Ser. Otd. Fiz.-Mat. Nauk, Vol. 7, No. 3 (1928).

Rode, E. Ya., Oxygen Compounds of Manganese, Izd. AN SSSR, Moscow (1952).

Rodina, A. G., Methods of Water Microbiology, Izd. Nauka (1965).

Rodt, V., "Aufklärungen zur Eisentrisulphidfrage und zur Entstehung des amorphen Eisensulphids," Z. angew. Chem., Vol. 29, No. 1 (1916).

Romankevich, E. A., "Organic matter in bottom deposits of the Pacific Ocean east of Kamchatka," Tr. Okeanografich. Komissii, Vol. 10, No. 2, Izd. AN SSSR (1960).

Romankevich, E. A., and Petrov, I. V., "Oxidation—reduction potential and pH of sediments in the northeastern part of the Pacific Ocean," Tr. Inst. Okeanologii, Vol. 14, Izd. AN SSSR (1961).

Romm, I. I., "Geochemical characteristics of recent deposits of the Taman' Peninsula," in: Recent Analogs of Oil-Bearing Facies, Gostoptekhizdat (1950).

Ronov, A. B., "Organic carbon in sedimentary rocks (in connection with oil content)," Geokhimiya, No. 5 (1958).

Rosenthal, G., "Versuche zur Darstellung von Markasit, Pirit, und Magnetkies aus wässrigen Losungen bei Zimmertemperatur," Heidelberger Beitr. Mineral. und Petrograph., Vol. 5, No. 2 (1956).

Rozhkova, E. V., Kuznetsova, É. G., and Vasil'eva, É. G., "Effect of bacterial process on the formation of epigenetic sulfide and other minerals in sedimentary rocks," Litolog. i Polezn. Iskop., No. 4 (1965).

Rubenchik, L. I., Sulfate-Reducing Bacteria, Izd. AN SSSR (1947).

Savich, V. G., "Physicochemical characteristics of basins and sediments of the Taman' Peninsula," in: Recent Analogs of Oil-Bearing Facies, Gostoptekhizdat (1950).

Savich, V. G., "Basic features of the oxidation—reduction state of recent marine sediments," in: Accumulation and Transformation of Organic Substances in Recent Marine Sediments, Gostoptekhizdat (1956).

Schmelck, L., On Oceanic Deposits, The Norwegian North Atlantic Expedition 1876–1878, Christiania (1882).

Semenovich, N. I., "Oxidation–reduction potential and pH in bottom deposits in Lake Ladoga," Dokl. Akad. Nauk SSSR, Vol. 151, No. 4 (1963).

Semenovich, N. I., "Oxidation–reduction potential and pH in bottom deposits in Lake Ladoga," in: Biological Basis of the Fishing Industry in the Internal Waters of the Baltic Region, Minsk (1964).

Serdobol'skii, I. P., "Methods of determining pH and oxidation–reduction potential in agrochemical investigations," in: Agrochemical Methods of Investigating Soils, Izd. Nauka (1965).

Serdyuchenko, D. P., "Chlorites, their chemical constitution and classification," Tr. Inst. Geol. Nauk, No. 140, Mineralogo-Geokhim. Seriya, No. 14, Izd. AN SSSR (1953).

Serpoyanu, G., "Penetration of Mediterranean waters into the basin of the Black Sea," Summary reports of the 2nd International Oceanographic Congress, Izd. Nauka (1966).

Shcherbakov, A. V., "Geochemical criteria of oxidation–reduction conditions in the underground hydrosphere," Sovetskaya Geologiya, No. 56 (1956).

Shcherbina, V. V., "Oxidation–reduction potential as applied to the study of mineral paragenesis," Dokl. Akad. Nauk, Vol. 22, No. 8 (1939).

Shcherbina, V. V., "Concentrations and dissemination of chemical elements in the earth's crust as a result of oxidation and reduction processes," Dokl. Akad. Nauk SSSR, Vol. 67, No. 3 (1949).

Shcherbina, V. V., Chemistry of Mineral-Forming Processes in Sedimentary Rocks. Questions on the Mineralogy of Sedimentary Rocks, Books 3 and 4, Izd. L'vovskogo Inst. (1956).

Shcherbina, V. V., The Role of Oxidation–Reduction Processes in the Formation of Ores [in Russian], Trudove V"rkhu Geologiyata ne B"lgariya, Ser. Geokhim. Mineralog. i Petrograf., Book 5, Sofia (1965).

Shishkina, O. V., "Oxidation–reduction potential in the upper ten-meter layer of the Quaternary rocks of the Black Sea," Dokl. Akad. Nauk SSSR, Vol. 139, No. 3 (1961).

Shneizer, G. M., and Zaidman, N. M., "Direct potentiometric determination of sulfides in water," Zavodskaya Laboratoriya, No. 3 (1965).

Sidorenko, M., "Petrographic data on recent deposits in the Khadzhibeiskii Estuary and the lithic composition of superficial sediments at the Kuyal'nitskii–Khadzhibeiskii overflow," Zap. Novoross. Obshch. Estestvoispytat, Vol. 24, No. 1 (1901).

Sillén, L. G., "Physical chemistry of sea water," in: Oceanography [Russian translation], Izd. Progress (1965).

Skinner, B. J., Erd, R. C., and Grimaldi, F. L., "Gregite, the thio spinel of iron; a new mineral," Amer. Mineralogist, Vol. 49, 5–6 (1964).

Skopintsev, B. A., "Distribution of hydrogen sulfide in the Black Sea," Meteorologiya i Gidrogeologiya, No. 7 (1953).

Skopintsev, B. A., "Study of the oxidation–reduction potential in waters of the Black Sea," Gidrokhim. Materialy, Vol. 27, Izd. AN SSSR (1957).

Skopintsev, B. A., Karpov, A. V., and Vershinina, O. A., "Experimental study of the formation and oxidation of hydrogen sulfide as observed in the Black Sea," Gidrokhim. Materialy, Vol. 21, Izd. AN SSSR (1961).

Skopintsev, B. A., and Smirnov, É. V., "Hydrogen sulfide in the deep waters of the open part of the Black Sea," Okeanologiya, Vol. 5, No. 6 (1965).

Slavin, P. S., "Types of oxidation–reduction conditions in the Cretaceous and Tertiary rocks of Turkmenia," Izv. Akad. Nauk SSSR, Ser. Geol., No. 7 (1961).

Sokolova, E. I., "Physicochemical investigation of ores and rocks of the Botomskii iron-ore deposit," in: Outlines of the Metallogeny of Sedimentary Rocks, Izd. AN SSSR (1961a).

Sokolova, E. I., "Physicochemical investigation of the iron-ore lake Punnus Yarvi," in: Outlines of the Metallogeny of Sedimentary Rocks, Izd. AN SSSR (1961b).

Sokolova, E. I., Physicochemical Investigations of Sedimentary Iron and Manganese Ores and Their Host Rocks, Izd. AN SSSR (1962).

Sokolova, G. A., "Distributional patterns of *Thiobacillus thioparus* in underground water," Mikrobiologiya, Vol. 30, No. 3 (1961).

Sokolova, G. A., and Sorokin, Yu. I., "Determination of intensity of bacterial reduction of sulfates in soils of the Gorki Reservoir by means of S^{35}," Dokl. Akad. Nauk SSSR, Vol. 118, No. 2 (1958).

Solomin, G. A., A Method of Determining Oxidation–Reduction Potential and pH in Sedimentary Rocks, Izd. Nauka (1964).

Sorokin, Yu. I., "Chemistry of the processes of aqueous reduction of sulfates," Tr. Inst. Mikrobiologii, No. 3, Izd. AN SSSR (1954).

Sorokin, Yu. I., "Experimental investigation of bacterial reduction of sulfates in the Black Sea," Mikrobiologiya, Vol. 31, No. 3 (1962).

Sorokin, Yu. I., "The role of dark bacterial assimilation of carbon dioxide in trophic basins," Mikrobiologiya, Vol. 33, No. 5 (1964).

Stashchuk, M. F., Lithic Aspects of Ancient Paleozoic Rocks along the Middle Dniester Region, Izd. AN UkrSSR, Kiev (1958).

Stashchuk, M. F., "Conditions for formation of iron minerals in marine sediments," in: Summary Reports of the Geochemical Conference Commemorating the 100th Birthday of Vernadskii, Izd. AN SSSR (1963).

Stashchuk (Staščuk), M. F., and Kropačeva, S. K., "Die Sulfide des dreiwertigen Eisens in rezenten Sedimenten und Sedimentgesteinen," Ber. Deutsch. Ges. Geol. Wiss., B. Miner. Lagerstätten, Vol. 14, 2, Berlin (1969).

Staschuk, M. F., Suprychev, V. D., and Khitraya, M. S., Mineralogy, Geochemistry, and Conditions of Formation of the Bottom Deposits in the Sivash, Izd. Naukova Dumka, Kiev (1964).

Strakhov, N. M., "Diagenesis of sediments and its meaning for sedimentary formation of ores," Izv. Akad. Nauk SSSR, Ser. Geol., No. 5 (1953).

Strakhov, N. M., Formation of Sediments in Modern Basins, Izd. AN SSSR (1954).

Strakhov, N. M., "Toward an understanding of diagenesis," in: Questions on the Mineralogy of Sedimentary Rocks, Books 3 and 4, L'vov (1956a).

Strakhov, N. M., "Some errors in method in studying chemical-biological sedimentation and diagenesis," Byull. MOIP, Vol. 31, No. 2 (1956b).

Strakhov, N. M., Principles of Lithogenesis, Vol. 1, Izd. AN SSSR (1960a).

Strakhov, N. M., Principles of Lithogenesis, Vol. 2, Izd. AN SSSR (1960b).

Strakhov, N. M., "Parageneses of authigenic minerals in sedimentary ores and factors determining them," Litolog. i Polezn. Iskop., No. 4 (1964).

Strakhov, N. M., and Zalmanzon, É. S., "Distribution of authigenic mineral forms of iron in sedimentary rocks and its significance in lithology," Izv. Akad. Nauk SSSR, Ser. Geol., No. 1 (1955).

Stukalova, M. M., "Determination of ferrous iron in silicates containing small amounts of sulfides," Materialy Vsesoyuz. Nauch.-issled. Geol. Inst., No. 6, Izd. AN SSSR (1947).

Sulin, V. A., and Varov, A. A., "The problem of scientific-research work in prospecting for oil in the Paleozoic rocks of the Ural–Volga region," Neftyanoe Khozyaistvo, Vol. 23, No. 7 (1932).

Taldykin, S. I., "Supergene magnetite in the Malka iron–chromium–nickel deposit," Sovetskoe Geologiya, No. 25 (1947).

Tauson, V. O., "Reduction of sulfates by bacteria in the presence of carbon dioxide," Mikrobiologiya, Vol. 1, No. 3 (1932).

Temple, K. L., "Syngenesis of sulfide ores: an evaluation of biochemical aspects," Econ. Geol., Vol. 59, No. 8 (1964).

Teodorovich, G. I., "Minerals of sedimentary rocks as indicators of physicochemical conditions," in: Questions on Mineralogy, Geochemistry, and Petrography, Izd. AN SSSR (1946).

Teodorovich, G. I., "Sedimentary geochemical facies," Byull. MOIP, Otd. Geol., Vol. 22, No. 1 (1947).

Teodorovich, G. I., "Sedimentary mineralogical–geochemical facies," in: Questions on the Mineralogy of Sedimentary Rocks, Books 3 and 4, Izd. L'vovskogo Univ. (1956).

Teodorovich, G. I., "Reducing hydrochemical conditions of ancient marine basins according to lithologic and geochemical features," in: Methods of Paleogeographic Investigation, Moscow (1964).

Thoulet, J., Etude Litologique des Fonds Recueillis dans les Parages de la Nouvelle Zemble, Campaque Arctique de 1907, Bruxelles (1910).

Tischendorf, G., and Ungethüm, H., "Über die Bedeutung des Reduktions–Oxidationspotentials (Eh) und der Wasserstoffionenkonzentration (pH) für Geochemie und Lagerstättenkunde," Geologie, Hft. 12 (1964).

Tischendorf, G., and Ungethüm, H., "Zur Anvendung von Eh–pH Beziehungen in den geologischen Praxis," Z. angew. Geol. Vol. 11, No. 2 (1965).

Tochilin, M. S., "Geochemistry of authigenic siderites," in: Questions on the Mineralogy of Sedimentary Rocks, Books 3 and 4, Izd. L'vovskogo Univ. (1956).

Treadwell, F. P., Kurzes Lehrbuch der Analytischen Chemie, Wien (1948–49).

Uspenskii, E. E., Energy of Life Processes; The Physicochemical Conditions of an Environment as a Basis of Microbiological Processes, Izd. AN SSSR (1936).

Vainbaum, S. Ya., "Geochemical facies of the Paleozoic of the Kuibyshev region in connection with questions of oil formation," Tr. Kuibyshevskogo Nauch.-issled. Inst. Neft. Promyshl., No. 1, Kuybyshev (1960a).

Vainbaum, S. Ya., "Reducing capacity as an index of epigenetic changes in rocks," Tr. Kuybyshevskogo Nauch.-issled. Inst. Neft. Promyshl., No. 1, Kuibyshev (1960b).

Vainbaum, S. Ya., and Il'inskaya, V. M., "Oxidation–reduction activity of Upper Devonian rocks in the Kuibyshev region," Tr. Kuybyshevskogo Nauch.-issled. Inst. Neft. Promyshl., No. 1, Kuibyshev (1960).

Veber, V., "From the Ermaka expedition in 1901," Zapiski Imp. SPb. Mineralogich. Obshch., 2nd Series, Pt. 46 (1908).

Volkov, I. I., "Determination of different forms of sulfur compounds in marine sediments," Tr. Inst. Okeanologii, Vol. 33, Izd. AN SSSR (1959).

Volkov, I. I., "Free hydrogen sulfide and some products of its transformation in sediments of the Black Sea," Tr. Inst. Okeanologii, Vol. 1, Izd. AN SSSR (1961a).

Volkov, I. I., "Iron sulfides, their interrelations and transformations in sediments of the Black Sea," Tr. Inst. Okeanologii, Vol. 1, Izd. AN SSSR (1961b).

Volkov, I. I., "Regularities in the formation and chemical composition of iron-sulfide concretions in deposits of the Black Sea," Tr. Inst. Okeanologii, Vol. 67, Izd. Nauka (1964).

Volkov, I. I., and Ostroumov, É. A., "Iron-sulfide concretions in deposits of the Black Sea," Dokl. Akad. Nauk SSSR, Vol. 116, No. 4 (1957).

Yagofarov, É. Kh., and Gorelova, T. L., "Authigenic minerals and some questions on the origin of Lower Carboniferous terrigenous rocks of the Kuma and Middle Volga region," Tr. Kuybyshevskogo Gos. Nauch.-Isled. Inst. Neftyanoi Promyshl., No. 11, Kuibyshev (1962).

Yamaguchi, S., and Katsurai, T., "Zur Bildung des ferromagnetischen Fe_3S_4, Kolloid. Zeit., Vol. 170 (1960).

Yanitskii, A. L., "The Georgiev deposit," in: The Khalilovo Deposit of Complex Ores, Tr. Inst. Geol. Nauk, No. 67, Ser. Rudn. Mestorozhd., No. 6, Izd. AN SSSR (1942).

Young, L., and Maw, G. A., The Metabolism of Sulfur Compounds, Methuen, London (1958).

Yurganov, N. N., "Joint geochemical investigations of sedimentary rocks for purposes of facies analysis," Tr. VNIGRI, No. 95, Gosgeoltekhizdat (1956a).

Yurganov, N. N., "Geochemical investigation of sedimentary rocks in the region of the Katangli oil field in northern Sakhalin," Tr. VNIGRI, No. 95, Gosgeoltekhizdat (1956b).

Yurkevich, I. A., Investigations by Facies – Geochemical Study of Sedimentary Rocks, Izd. AN SSSR (1958).

Yurkevich, I. A., "Facies–geochemical investigations in describing source beds of petroleum," in Geochemistry of Oil and Oil Deposits, Izd. AN SSSR (1962).

Zavodnov, S. S., Carbonate and Sulfide Equilibrium in Mineral Waters, Gidrometeoizdat, Leningrad (1965).

Zelenskii, N. D. "Hydrogen-sulfide fermentation in the Black Sea and the Odessa estuaries," Zh. Russk. Fiz.-khim. Obshch., T. Khim., Vol. 25, No. 5 (1893).

Zelenskii, N. D., and Brusilovskii, E. M., "Hydrogen-sulfide fermentation in the Black Sea," Yuzhnorussk. Meditsinsk. Gazeta, No. 18–19 (1893).

Zheleznova, A. A., and Shishkina, O. V., "Oxidation–reduction potential in active reaction of sediments in the northern part of the Indian Ocean," Tr. Inst. Okeanologii AN SSSR, Vol. 64, Izd. AN SSSR (1964).

Zobell, C. E., "Studies on redox potential of marine sediments," Am. Assoc. Petroleum Geologists Bull., Vol. 4 (1946).

LITERATURE CITED